TRANSITIONS TO ALTERNATIVE TRANSPORTATION TECHNOLOGIES — A FOCUS ON HYDROGEN

Committee on Assessment of Resource Needs for Fuel Cell and Hydrogen Technologies

Board on Energy and Environmental Systems
Division on Engineering and Physical Sciences

NATIONAL RESEARCH COUNCIL
OF THE NATIONAL ACADEMIES

THE NATIONAL ACADEMIES PRESS
Washington, D.C.
www.nap.edu

THE NATIONAL ACADEMIES PRESS • 500 Fifth Street, N.W. • Washington, DC 20001

NOTICE: The project that is the subject of this report was approved by the Governing Board of the National Research Council, whose members are drawn from the councils of the National Academy of Sciences, the National Academy of Engineering, and the Institute of Medicine. The members of the committee responsible for the report were chosen for their special competences and with regard for appropriate balance.

This study was supported by Contract DE-AT01-06EE11206, TO#18, Subtask 3 between the National Academy of Sciences and the U.S. Department of Energy. Any opinions, findings, conclusions, or recommendations expressed in this publication are those of the author(s) and do not necessarily reflect the views of the organizations or agencies that provided support for the project.

Cover: Photos of a fuel cell vehicle and hydrogen fueling station courtesy of the U.S. Department of Energy.

Library of Congress Cataloging-in-Publication Data

National Research Council (U.S.). Committee on Assessment of Resource Needs for Fuel Cell and Hydrogen Technologies.
Transitions to alternative transportation technologies : a focus on hydrogen / Committee on Assessment of Resource Needs for Fuel Cell and Hydrogen Technologies, Board on Energy and Environmental Systems, Division on Engineering and Physical Sciences.
p. cm.
Includes bibliographical references.
ISBN 978-0-309-12100-2 (pbk.) — ISBN 978-0-309-12101-9 (pdf) 1. Fuel cell vehicles—Research—Government policy—United States. 2. Hydrogen as fuel—Research—Government policy—United States. I. Title.
TL221.13.N38 2008
629.22′9—dc22

2008036579

Available in limited supply from:

Board on Energy and Environmental Systems
National Research Council
500 Fifth Street, N.W.
Keck W934
Washington, DC 20001
(202) 334-3344

Additional copies for sale from:

The National Academies Press
500 Fifth Street, N.W.
Lockbox 285
Washington, DC 20055
(800) 624-6242 or (202) 334-3313
(in the Washington metropolitan area)
Internet: http://www.nap.edu

Printed in the United States of America

THE NATIONAL ACADEMIES

Advisers to the Nation on Science, Engineering, and Medicine

The **National Academy of Sciences** is a private, nonprofit, self-perpetuating society of distinguished scholars engaged in scientific and engineering research, dedicated to the furtherance of science and technology and to their use for the general welfare. Upon the authority of the charter granted to it by the Congress in 1863, the Academy has a mandate that requires it to advise the federal government on scientific and technical matters. Dr. Ralph J. Cicerone is president of the National Academy of Sciences.

The **National Academy of Engineering** was established in 1964, under the charter of the National Academy of Sciences, as a parallel organization of outstanding engineers. It is autonomous in its administration and in the selection of its members, sharing with the National Academy of Sciences the responsibility for advising the federal government. The National Academy of Engineering also sponsors engineering programs aimed at meeting national needs, encourages education and research, and recognizes the superior achievements of engineers. Dr. Charles M. Vest is president of the National Academy of Engineering.

The **Institute of Medicine** was established in 1970 by the National Academy of Sciences to secure the services of eminent members of appropriate professions in the examination of policy matters pertaining to the health of the public. The Institute acts under the responsibility given to the National Academy of Sciences by its congressional charter to be an adviser to the federal government and, upon its own initiative, to identify issues of medical care, research, and education. Dr. Harvey V. Fineberg is president of the Institute of Medicine.

The **National Research Council** was organized by the National Academy of Sciences in 1916 to associate the broad community of science and technology with the Academy's purposes of furthering knowledge and advising the federal government. Functioning in accordance with general policies determined by the Academy, the Council has become the principal operating agency of both the National Academy of Sciences and the National Academy of Engineering in providing services to the government, the public, and the scientific and engineering communities. The Council is administered jointly by both Academies and the Institute of Medicine. Dr. Ralph J. Cicerone and Dr. Charles M. Vest are chair and vice chair, respectively, of the National Research Council.

www.national-academies.org

COMMITTEE ON ASSESSMENT OF RESOURCE NEEDS FOR FUEL CELL AND HYDROGEN TECHNOLOGIES

MICHAEL P. RAMAGE, NAE,[1] ExxonMobil Research and Engineering Company (retired), *Chair*
RAKESH AGRAWAL, NAE, Purdue University
DAVID L. BODDE, Clemson University
DAVID FRIEDMAN, Union of Concerned Scientists
SUSAN FUHS, Conundrum Consulting
JUDI GREENWALD, Pew Center on Global Climate Change
ROBERT L. HIRSCH, Management Information Services, Inc.
JAMES R. KATZER, NAE, Massachusetts Institute of Technology
GENE NEMANICH, ChevronTexaco Technology Ventures (retired)
JOAN OGDEN, University of California, Davis
LAWRENCE T. PAPAY, NAE, Science Applications International Corporation (retired)
IAN W.H. PARRY, Resources for the Future
WILLIAM F. POWERS, NAE, Ford Motor Company (retired)
EDWARD S. RUBIN, Carnegie Mellon University
ROBERT W. SHAW, JR. Aretê Corporation
ARNOLD F. STANCELL, NAE, Georgia Institute of Technology
TONY WU, Southern Company

Staff

Board on Energy and Environmental Systems

ALAN CRANE, Study Director
MATT BOWEN, Senior Program Associate
DUNCAN BROWN, Senior Program Officer
JAMES ZUCCHETTO, Director, BEES

National Academy of Engineering Program Office

PENELOPE GIBBS, Senior Program Associate

[1]NAE, National Academy of Engineering.

BOARD ON ENERGY AND ENVIRONMENTAL SYSTEMS

Staff

[1]NAE, National Academy of Engineering.
[2]NAS, National Academy of Sciences.

Preface

Hydrogen is a potential replacement fuel for gasoline in light-duty vehicles. Hydrogen fuel cell vehicles could alleviate the nation's dependence on oil and reduce U.S. emissions of carbon dioxide, the major greenhouse gas.

Industry- and government-sponsored research programs have made very impressive technical progress over the past several years, and several companies are currently introducing pre-commercial vehicles and hydrogen fueling stations in limited markets. The introduction of fuel cell vehicles into the light-duty vehicle fleet is much closer to reality than when the National Research Council (NRC) last examined the technology in 2004.

However, to achieve wide hydrogen vehicle penetration, further technological advances are required for commercial viability, and vehicle manufacturer and hydrogen supplier activities must be coordinated. In particular, costs must be reduced, new automotive manufacturing technologies commercialized, and adequate supplies of hydrogen produced and made available to motorists. These efforts will require considerable resources, especially federal and private sector funding.

This report estimates the resources that will be needed to bring hydrogen fuel cell vehicles to the point of competitive self-sustainability in the marketplace. It also estimates the impact on oil consumption and carbon dioxide emissions as fuel cell vehicles become a large fraction of the light-duty vehicle fleet. The study was requested by the U.S. Congress in the Energy Policy Act of 2005 and contracted for by the U.S. Department of Energy.

I greatly appreciate the efforts made by the many highly qualified experts on the committee. The committee operated under the auspices of the NRC Board on Energy and Environmental Systems and is grateful for the able assistance of James Zucchetto, Alan Crane, and Duncan Brown of the NRC staff, and of Penelope Gibbs of the National Academy of Engineering Program Office staff.

Michael P. Ramage, *Chair*
Committee on Assessment of Resource Needs
for Fuel Cell and Hydrogen Technologies

Acknowledgments

The Committee on Assessment of Resource Needs for Fuel Cell and Hydrogen Technologies is grateful to the many individuals who contributed their time and efforts to this National Academies' National Research Council (NRC) study. The presentations at committee meetings provided valuable information and insights that enhanced the committee's understanding of the technologies and barriers involved. The committee thanks the following individuals who provided briefings:

Phillip Baxley, Shell Hydrogen LLC
Jon Bereisa, General Motors Corporation
K.G. Duleep, ICF
Catherine Dunwoody, California Fuel Cell Partnership
Kelly Fletcher, GE Research
David Greene, Oak Ridge National Laboratory
Sig Gronich, U.S. Department of Energy (DOE)
Knut Harg, Norsk-Hydro
Richard Hess, Idaho National Laboratory
Brian James, Directed Technologies, Inc.
Bryan Jenkins, University of California, Davis
Timothy Johnson, Environmental Protection Agency
Fred Joseck, U.S. DOE
Arthur Katsaros, Air Products
Taiyo Kawai, Toyota Motor Company
Ben Knight, Honda Motor Corporation
Johanna Levene, National Renewable Energy Laboratory
Margaret Mann, National Renewable Energy Laboratory
Fred Maples, Energy Information Administration
Lowell Miller, U.S. DOE
JoAnn Milliken, U.S. DOE
Joan Ogden, University of California, Davis
Mark Paster, U.S. DOE
Steve Plotkin, Argonne National Laboratory
Dan Rastler, Electric Power Research Institute
Bill Reinert, Toyota Motor Sales, USA, Inc.
Robert Rose, U.S. Fuel Cell Council
Robert W. Shaw, Aretê Corporation
Frances Wood, On Location, Inc.

This report was reviewed in draft form by individuals chosen for their diverse perspectives and technical expertise, in accordance with procedures approved by the NRC's Report Review Committee. The purpose of the independent review is to provide candid and critical comments that will assist the institution in making its published report as sound as possible and to ensure that the report meets institutional standards for objectivity, evidence, and responsiveness to the study charge. The review comments and draft manuscript remain confidential to protect the integrity of the deliberative process. We wish to thank the following individuals for their review of this report:

Allen J. Bard (NAS), University of Texas, Austin
Phillip Baxley, Shell Hydrogen LLC
Deborah Bleviss, Consultant
Andrew Brown, Jr. (NAE), Delphi Corporation
Douglas Chapin (NAE), MPR Associates, Inc.
Robert Epperly, Consultant
Paul Gilbert (NAE), Parsons Brinckerhoff, Inc. (retired)
Trevor Jones (NAE), ElectroSonics Medical, Inc.
Vernon P. Roan, University of Florida (emeritus)
James Sweeney, Stanford University
G. David Tilman (NAS), University of Minnesota.

Although the reviewers listed above have provided many constructive comments and suggestions, they were not asked to endorse the conclusions or recommendations, nor did they see the final draft of the report before its release. The review of this report was overseen by Elisabeth M. Drake, Massachusetts Institute of Technology, and Maxine L. Savitz, Honeywell, Inc. (retired). Appointed by the National Research Council, they were responsible for making certain that an independent examination of this report was carried out in accordance with institutional procedures and that all review comments were carefully considered. Responsibility for the final content of this report rests entirely with the authoring committee and the institution.

Contents

Tables, Figures, and Boxes

TABLES

FIGURES

BOXES

Abstract

In response to a congressional request in the Energy Policy Act of 2005, this National Research Council (NRC) study estimated the maximum practicable number of hydrogen fuel cell vehicles (HFCVs) that could be deployed in the United States by 2020 and beyond, together with the investments, time, and government actions needed to carry out this transition. The study determined the consequent reductions in U.S. oil consumption and emissions of carbon dioxide (CO_2)—the main greenhouse gas linked to global climate change—that could be expected. It then compared those reductions with the potential impact that the use of alternative vehicle technologies and biofuels might have on oil consumption and CO_2 emissions.

The NRC's Committee on Assessment of Resource Needs for Fuel Cell and Hydrogen Technologies concluded that the maximum practical number of HFCVs that could be operating in 2020 would be approximately 2 million in a fleet of 280 million light-duty vehicles. The number of HFCVs could grow rapidly thereafter to about 25 million by 2030. Rather than a prediction of the future by the committee, this is a scenario based on the committee's estimate of the maximum penetration rate, assuming that technical goals are met, that consumers readily accept HFCVs, and that policy instruments are in place to drive the introduction of hydrogen fuel and fuel cell vehicles through the market transition period.

The use of HFCVs can achieve large and sustained reductions in U.S. oil consumption and CO_2 emissions, but several decades will be needed to realize these potential long-term benefits. Considerable progress is still required toward improving fuel cell costs and durability, as well as on-board hydrogen storage. The substantial financial commitments and technical progress made in recent years by the automotive industry, private entrepreneurs, and the U.S. Department of Energy (DOE) suggest that HFCVs and hydrogen production technologies could be ready for commercialization in the 2015-2020 time frame. Such vehicles are not likely to be cost-competitive until after 2020, but by 2050 HFCVs could account for more than 80 percent of new vehicles entering the fleet.

An accelerated transition to HFCVs would require that automobile manufacturers ramp up production of fuel cell vehicles even while they cost much more than conventional vehicles, and that investments be made to build and operate hydrogen fueling stations even while the market for hydrogen is very limited. Substantial government actions and assistance would therefore be needed to support such a transition to HFCVs in the 2020 time frame, even with good technical progress on fuel cell and hydrogen production technologies. Substantial and sustained research and development (R&D) programs also are required to further reduce the costs of fuel cell vehicles and hydrogen after 2020.

The committee estimated the government cost to support a transition to hydrogen fuel cell vehicles as being roughly \$55 billion from 2008 to 2023 (when fuel cell vehicles would become competitive with gasoline-powered vehicles). This funding includes a substantial R&D program (\$5 billion), support for the demonstration and deployment of the vehicles while they are more expensive than conventional vehicles (\$40 billion), and support for the production of hydrogen (\$10 billion). Private industry would be investing far more, about \$145 billion for R&D, vehicle manufacturing, and hydrogen infrastructure over the same period.

Current U.S. government expenditures, largely for R&D, are about \$300 million per year, primarily by the U.S. Department of Energy. If 2 million HFCVs are to be on the road by 2020, R&D funding may have to be increased by as much as 20 percent over the next several years. Annual government expenditures will have to be much higher to support the commercial introduction of HFCVs, about \$3 billion in 2015 and increasing to \$8 billion in 2023.

Potential synergies between the transportation sector and the electric power sector may help reduce the cost of hydrogen. In the near term, electrolysis of water can provide hydrogen in areas where natural gas or other sources are unavailable. In the longer term (after 2025), co-generation of low-carbon hydrogen and electricity in gasification-based energy plants may be an option.

The main advantage of a transition to HFCVs is the potential for reducing the use of oil and emissions of CO_2. Although hydrogen could not replace much gasoline before 2025, the 25 years after that would see a dramatic decline in the use of gasoline in the light-duty vehicle fleet to about one-third of current projections, if the assumptions of the maximum practical case are met. Emissions of CO_2 will decline almost as much if hydrogen is produced with carbon capture and sequestration or from nonfossil sources.

The committee also found that alternatives such as improved fuel economy for conventional vehicles, increased penetration of hybrid vehicles, and biomass-derived fuels could deliver significantly greater reductions in U.S. oil use and CO_2 emissions than could use of HFCVs over the next two decades, but that the longer-term benefits of such approaches were likely to grow at a smaller rate thereafter, even with continued technological improvements, whereas hydrogen offers greater longer-term potential. Thus, as estimated by the committee, the greatest benefits will come from a portfolio of R&D technologies that would allow the United States to achieve deep reductions in oil use, nearly 100 percent by 2050 for the light-duty vehicle fleet. Achieving this goal, however, will require significant new energy security and environmental policy actions in addition to technological developments. Although broad policies aimed at reducing oil use and CO_2 emissions will be useful, they are unlikely to be adequate to facilitate the rapid introduction of HFCVs. A competitive and self-sustaining HFCV fleet is possible in the long term but will require hydrogen-specific policies in the nearer term. These policies must be substantial and durable in order to assure industry that the necessary long-term investments can be made safely.

Summary

BACKGROUND AND SCOPE OF STUDY

In 2005, drawn by the potential for hydrogen-fueled vehicles to achieve large reductions in U.S. oil imports and emissions of carbon dioxide (CO_2)—the main greenhouse gas linked to global climate change—Congress requested that the National Research Council (NRC) assess what resources would be required for a transition in the U.S. light-duty vehicle fleet to hydrogen fuel cell vehicles (HFCVs) by 2020. Specifically, Section 1825 of the Energy Policy Act of 2005 stated: "The National Academy of Sciences' National Research Council will appoint a committee to carry out a study of fuel cell technologies that provides a budget roadmap (e.g., what investments in R&D, demonstrations, skilled people, [and] infrastructure will be required) for the development of fuel cell technologies and the transition from petroleum to hydrogen in a significant percentage of the vehicles sold by 2020."

In 2007, the NRC formed the Committee on Assessment of Resource Needs for Fuel Cell and Hydrogen Technologies. The statement of task for this study requested that the committee:

- Establish as a goal the *maximum practicable number* of vehicles that can be fueled by hydrogen by 2020;
- Determine the *funding*, public and private, to reach that goal;
- Determine the *government actions* required to achieve the goal;
- Consider the role that hydrogen's use in *stationary electric power* applications will play in stimulating the transition to hydrogen-fueled hybrid electric vehicles;
- Consider whether *other technologies* could achieve significant CO_2 and oil reductions by 2020; and
- Establish a *budget roadmap* to achieve the goal.

Early in its deliberations, the committee noted that the congressional focus on impact by 2020 implied an urgency for actions to reduce U.S. oil imports and CO_2 emissions. But on the basis of recent studies, including the NRC's 2004 report *The Hydrogen Economy* (NRC, 2004), the committee concluded that it would not be feasible to have enough hydrogen vehicles on the road by 2020 to significantly affect CO_2 emissions and oil use, although hydrogen could have a substantial impact in the longer run. Thus, the committee extended the time period of its study to 2050 and estimated the technical readiness and potential impacts of HFCVs at 2020, 2035, and 2050.

The emphasis throughout the study, as set forth in the statement of task, is on the *maximum practicable number* of HFCVs, or, as used in this study, the maximum practicable penetration rate (MPR) of HFCVs achievable in the 2008 to 2050 time frame. Rather than a prediction of the future, the committee developed a scenario based on its estimate of the maximum practicable penetration rate, assuming that technical goals are met, that consumers readily accept HFCVs, and that policy instruments are in place to drive the introduction of hydrogen fuel and fuel cell vehicles through the market transition period.

In keeping with its statement of task, the committee also considered whether other technologies might achieve significantly greater reductions in oil imports and CO_2 emissions than HFCVs over the next several decades. After considering a range of alternative technologies and the budget constraints of the study, the committee chose to quantitatively evaluate, using the MPR approach, one alternative fuel and one alternative vehicle option, namely, (1) fuels derived from biomass (in light of the increased emphasis on this option in the United States) and (2) evolutionary improvements in internal combustion engines (ICEs) and hybrid electric vehicles (HEVs) (in light of the potential of these technologies to increase vehicle efficiency in the short term). These alternative fuel and vehicle technologies also will be needed through 2020 to meet the significantly higher fuel economy standards required by the Energy Independence and Security Act (EISA) of 2007.

Two other alternative technologies that are likely to contribute to improved U.S. fuel economy are electric vehicles, including plug-in hybrid electric vehicles (PHEVs), and diesel engines in light-duty vehicles. However, these options were not explicitly evaluated by the committee, both because of resource limitations and because uncertainties in the future costs and consumer acceptance of these technologies were judged to be too great for the committee to have confidence in any assumed penetration rates. For all technologies, the study was restricted to light-duty vehicles (automobiles and light trucks), which represent the bulk of the U.S. vehicle market.

As a benchmark for evaluating the ability of HFCVs and other technologies to reduce oil imports and CO_2 emissions, the committee developed a reference case scenario based on the high-price case of the Energy Information Administration's (EIA's) *Annual Energy Outlook 2008* (EIA, 2008). This scenario included a significant increase in fuel economy standards as required by the Energy Independence and Security Act of 2007. The committee used the EIA report to estimate factors such as the future number of vehicles of different types on the road, as well as reference case fuel economies and gasoline prices. The committee recognizes, of course, that the projections presented in this study—like any other—contain significant uncertainties and unknowns because of changes that are likely to occur over the next several decades.

SYNOPSIS OF STUDY RESULTS

The substantial financial commitments and technical progress made by the automotive industry, private entrepreneurs, and the U.S. Department of Energy (DOE) in hydrogen fuel cell and hydrogen production technologies suggest the potential for progress to the point that commercial HFCVs could be introduced in the United States in 2015-2020. However, these vehicles are unlikely to be cost-competitive until several years after 2020 even if the maximum practicable number is reached. It will thus require substantial government action (e.g., subsidies and enactment of regulations), plus continued support for research, development, and demonstration (RD&D), to move HFCVs into the market in sufficient numbers to reduce costs and make the technology self-supporting in the marketplace. Nevertheless, the committee's analysis also showed that the long-term promise of HFCVs in reducing oil consumption and CO_2 emissions is significant, and potentially greater than that of other nearer-term alternatives. Although it was not asked to make a formal analysis of the value of policies to support hydrogen, the committee believes that, in view of the potential risks posed by oil supply disruptions and increasing CO_2 emissions from oil use, the magnitude of the potential benefits justifies sustained government support of hydrogen vehicle development as part of a portfolio of options to address these serious national problems.

The committee's analysis indicated that over the next two decades, a combination of improved conventional and hybrid vehicle fuel economy, together with increased use of biomass-derived fuels (known generically as biofuels)—and with sufficient market conditions and policies in place—could deliver substantial reductions in U.S. oil use and CO_2 emissions. While HFCVs are unlikely to deliver significant benefits in this period, eventually they can do much better. Thus HFCVs are not direct competitors with other options that are able to deliver more immediate environmental and fuel use benefits. Instead, if employed with these options, collectively they can achieve dramatic, long-term reductions in oil use and CO_2 emissions—benefits that could continue to grow beyond the 2030-2050 time frame. Achieving significant benefits of this kind, however, will require additional policy measures to promote the early introduction of fuel cell vehicles and to ensure that hydrogen is produced in ways that do not add to the CO_2 burden.

CONCLUSION 1: A portfolio of technologies including hydrogen fuel cell vehicles, improved efficiency of conventional vehicles, hybrids, and use of biofuels—in conjunction with required new policy drivers—has the potential to nearly eliminate gasoline use in light-duty vehicles by the middle of this century, while reducing fleet greenhouse gas emissions to less than 20 percent of current levels. This portfolio approach provides a hedge against potential shortfalls in any one technological approach and improves the probability that the United States can meet its energy and environmental goals. Other technologies also may hold promise as part of a portfolio, but further study is required to assess their potential impacts. See Chapter 9.

CONCLUSION 2: Sustained, substantial, and aggressive energy security and environmental policy interventions will be needed to ensure marketplace success for oil-saving and greenhouse-gas-reducing technologies, including hydrogen fuel cell vehicles. See Chapter 8.

TECHNOLOGY ASSESSMENT

To develop the maximum practicable number scenario and associated budget roadmap, the committee assessed the technical progress and future challenges for (1) hydrogen fuel cell vehicles and (2) hydrogen production and delivery systems.

Hydrogen Fuel Cell Vehicles

Concentrated efforts by private companies, together with the U.S. FreedomCAR Fuel Partnership (FCFP) and other government-supported programs around the world, have resulted in significant progress toward a commercially viable hydrogen fuel cell vehicle since the publication in

2004 of *The Hydrogen Economy* (NRC, 2004). Fuel cell costs have been reduced significantly over the past 4-5 years. Costs projected for high-volume (500,000 units per year) automotive fuel cell production are approximately $100/kW[1] for relatively proven technologies and $67/kW for newer laboratory-based technologies, compared to the DOE-FCFP commercialization goal for 2015 of $30/kW. The cost of platinum is approximately 57 percent of the fuel cell stack costs and represents the greatest challenge to further cost reductions. Fuel cell stack life has increased to approximately 2,000 hours compared to the DOE-FCFP 2015 goal of 5,000 hours. Recently, new failure modes have been identified, particularly platinum dissolution into the carbon electrodes. However, focused research to ameliorate these failure modes, together with recent advances in electrode and membrane technology, should further reduce costs and increase stack life. Onboard hydrogen storage to achieve a 300-mile driving range has been the most difficult technical challenge. Identification of solid storage materials to achieve the DOE-FCFP 2015 goals, including the cost goal of $2/kWh, is in the research stage. It is not clear at this time whether a suitable material will be identified that can meet these goals and timing targets. However, in order to achieve the desired driving range between refueling stops, the industry is prepared to use more expensive high-pressure hydrogen storage tanks that consume more space and add to vehicle weight while research progresses toward a more commercially viable hydrogen storage material. HFCV fuel economy is currently about 50-55 miles per gallon of gasoline equivalent (mpgge) for a midsize vehicle with an 80 kW fuel cell. The DOE target of 60 percent efficiency for 2015 corresponds to 80 mpgge for the typical vehicle considered in this report.

In summary, paths forward have been identified for further reducing hydrogen fuel cell costs while increasing durability and fuel economy. Based on its technical assessment, the committee concluded that under the maximum practicable number of vehicles scenario, a significant market transition to HFCVs could start around 2015 if supported by strong government policies to drive early growth, even if DOE technology targets are not fully realized. The analysis of potential reductions in oil use and CO_2 emissions, discussed below, is based on the committee's understanding of the current technical status of fuel cell vehicles and the potential for improvements over the next several decades.

CONCLUSION 3: Lower-cost, durable fuel cell systems for light-duty vehicles are likely to be increasingly available over the next 5-10 years, and, if supported by strong government policies, commercialization and growth of HFCVs could get underway by 2015, even though all DOE targets for HFCVs may not be fully realized. See Chapter 3.

[1]One kilowatt (kW) is equal to 1.34 horsepower. A kilowatt-hour (kWh) is the work done by 1 kW operating for an hour.

Hydrogen Production Systems

To develop a budget roadmap for the maximum practicable number scenario, the committee also evaluated the status of hydrogen production technologies. In the committee's judgment, the three hydrogen production technologies that have the highest likelihood of commercial viability in the 2015-2035 time frame are (1) distributed steam methane reformation (DSMR) using natural gas as a feedstock for on-site production at a refueling station; (2) centralized hydrogen production from coal gasification with carbon capture and sequestration (CCS); and (3) centralized production from biomass gasification. Carbon-free hydrogen produced using advanced high-temperature nuclear reactors for electrolysis or thermochemical splitting of water might also be possible in this time frame, but the timetable and costs for development and commercialization of advanced (Generation IV) nuclear technology are difficult to estimate. Electrolysis using electricity from the grid may be useful in certain circumstances but is likely to be more expensive than DSMR in most cases. Hydrogen produced from renewable energy sources, such as electricity generated by solar and wind energy, could be viable if these technologies become more extensively deployed and their costs decline significantly (especially solar energy systems). Direct energy conversion systems using photoelectrochemical or photobiological technologies also can be significant long-term contributors to a hydrogen economy if associated technical hurdles can be overcome.

Based on its assessment of hydrogen supply options, the committee concluded that in a *maximum practicable* scenario:

- DSMR technology can be commercially available in sufficient quantities to fuel HFCVs at the DOE-FCFP cost goal of $3.00/kg hydrogen in 2015, when a transition to fuel cell vehicles might begin. Because a kilogram of hydrogen has about the same energy as a gallon of gasoline but is used twice as efficiently in a fuel cell, $3.00/kg is equivalent to $1.50/gallon. DSMR is likely to be more economical than alternative on-site technologies (such as electrolysis) and will be sufficient to fuel HFCVs through about 2025. Even though DSMR generates CO_2 that is not captured, the report *The Hydrogen Economy* (NRC, 2004) showed that well-to-wheels CO_2 from DSMR-HFCV could be less than half that from conventional gasoline-powered vehicles. The quantity of natural gas used through 2025 (approximately a 2 percent increase in projected demand, based on EIA data) should not be large enough to dramatically affect natural gas prices. Delivered hydrogen costs at the outset of the transition will be high because of the underutilization of production equipment with a small number of HFCVs on the road. However, by 2016 the cost of hydrogen is projected to equal that of gasoline on a dollars-per-mile-of-travel basis. The future price of natural gas is the largest determinant of the viability of this option.

• Hydrogen production via coal gasification plus carbon capture and storage (CCS), with pipeline delivery of hydrogen to refueling stations, could be commercially viable and available by 2025. If coal is used, CCS is necessary to achieve low-carbon hydrogen production. Coal gasification and associated carbon capture technologies are already used commercially to produce hydrogen (albeit on a much smaller scale than natural gas reforming), but they have not yet been integrated with carbon sequestration. Nor has deep geological sequestration of CO_2 yet been demonstrated in the United States at the scale envisioned for a commercial hydrogen plant. Thus, the greatest challenge to coal-based hydrogen production is demonstrating the capacity and long-term storage capabilities for geological sequestration of CO_2 in deep saline aquifers. However, as part of the DOE Carbon Sequestration Program, there are several industrial-scale, well-monitored demonstration projects currently planned to address issues of commercial viability over the next several years, and three large-scale geological sequestration projects (1 million tons of CO_2 per year each) have been operating successfully in other parts of the world for the past 4-12 years.

• Hydrogen from biomass gasification technology also is advancing and could be competitive by the mid to late 2020s for centralized production with potentially low carbon impacts. Several scale-up projects are now under way. Carbon dioxide from biomass gasification also potentially can be captured and stored to yield "negative emissions" of CO_2. While this hydrogen technology currently is not as well developed as the two outlined above, the committee included it as a renewable hydrogen source. Recent advances indicate that the cost could approach $3.00/kg hydrogen with continued technology progress.

The main challenges to deploying these technologies are (1) developing the technical capacity and regulatory framework to permit and safely sequester CO_2; (2) developing the technical, economic, and environmental knowledge needed to support large-volume biomass production and transportation to central gasification facilities; and (3) establishing public confidence in procedures for efficient permitting and installation of hydrogen production, distribution, and refueling systems. Safety issues, both real and perceived, along with the creation of appropriate codes and standards, are significant barriers to the introduction of HFCVs and the development of a hydrogen refueling infrastructure and pose a significant risk to achieving the maximum practicable penetration rate for hydrogen vehicles. Under the *maximum percentage practicable* scenario, the committee assumes that these hurdles will be overcome.

CONCLUSION 4: If appropriate policies are adopted to accelerate the introduction of hydrogen and HFCVs, hydrogen from distributed technologies can be provided at reasonable cost to initiate the maximum practicable case. If technical targets for central production technologies are met, lower-cost hydrogen should be available to fuel HFCVs in the latter part of the time frame considered in this study. Additional policy measures are required to achieve low-carbon hydrogen production in order to significantly reduce CO_2 emissions from central coal-based plants. See Chapter 3.

MAXIMUM PRACTICABLE DEPLOYMENT OF FUEL CELL VEHICLES

A key task of this study was to "establish . . . the maximum practicable number of vehicles that can be fueled by hydrogen by 2020." Based on that number of vehicles, the committee was then to assess the impact on oil use and reduction of CO_2 emissions; determine the costs and budgets needed to implement a program of this magnitude; and outline government actions that might be necessary to achieve it. The committee concluded that a goal for the maximum practicable number of vehicles would depend strongly on a host of factors related to future rates of technical progress in both fuel cell vehicles and hydrogen production, as well as future policy actions and consumer preferences. Given the many uncertainties, the committee developed three scenarios to reflect a range of possible HFCV penetration rates, to help guide its judgment in this task:

1. *Hydrogen Success (Case 1).* This scenario assumes that hydrogen and fuel cell developments meet slightly reduced performance and cost goals compared with those established by DOE and the FreedomCar Fuel Partnership Program (as explained in Chapter 3). This rapid-growth case corresponds to a scenario recently developed by DOE to 2025 (Gronich, 2007), and extended by the committee to 2050. By 2050, 80 percent of new vehicles sold are assumed to be HFCVs (see Figure 6.2 in Chapter 6). This is consistent with other recent modeling studies (Greene et al., 2007).

2. *Accelerated Success (Case 1a).* This variant assumes a far more aggressive program that doubles the rate of introduction of HFCVs compared with Case 1, while meeting the same performance and cost goals. Very aggressive national policy measures would be needed to drive this accelerated case—for example, much more stringent CAFE standards in response to severe curtailments in oil supplies, or stringent limits on CO_2 emissions in response to major new concerns about climate change.

3. *Partial Success (Case 1b).* The third case assumes shortfalls and delays in reaching the performance and cost goals of Case 1, with penetration rates remaining at historical rates for hybrids, supported by government subsidies.

Details of all scenario assumptions are elaborated in Chapter 6 of this report. The committee used these scenarios to establish its estimate of the maximum practicable number of vehicles. Analyses were conducted using a model developed by committee members and based on other recent modeling studies and analyses (Chapter 6 provides all details). Results of the analysis include the timing and magnitude of costs for HFCVs and associated infrastructure deployment, as well as the resulting impacts on oil use and CO_2 emissions relative to a base case scenario also defined by the committee.

In the committee's view, the Hydrogen Success case (Case 1) best represents the maximum practicable number of vehicles that could be fueled by hydrogen by 2020. The Accelerated Success case (Case 1a), while achieving greater reductions in oil consumption and CO_2 emissions, had substantially higher costs and greater technical risks. The Partial Success case (Case 1b), which was more pessimistic about achievable rates of technical progress, was higher in total cost than Case 1 because it took longer to achieve significant HFCV penetration and commercial viability, and Case 1b did not have a significant impact on oil and CO_2 reductions over the next few decades.

Achieving the Hydrogen Success case, however, would be challenging—requiring significant continued technical progress, consumer acceptance, and policies to achieve market penetration of HFCVs during the early transition period. Thus, it is by no means a "sure thing."

For these reasons, *it should be understood throughout this report that the maximum practicable number of vehicles estimated by the committee in response to the statement of task does not represent the committee's view of the "probable" or "most likely" number of HFCVs on the road by 2020 and beyond.* Rather, the estimate of maximum practicable number reflects a judgment about the ability to achieve the performance and cost goals required for market competitiveness, grounded in historically observed rates of market penetration for other new vehicle concepts and technologies, which have not faced the difficulties associated with establishing a major new fuel infrastructure. *Although more aggressive scenarios can be envisioned under certain circumstances, in general, the Hydrogen Success case should be viewed as an optimistic estimate of what might be possible—not a forecast of what is likely or probable.*

Figure S.1 shows the number of hydrogen fuel cell vehicles in the U.S. fleet and the percentage of new vehicles sold over time for the Hydrogen Success case. Trends for HFCVs are compared with those for conventional gasoline-powered vehicles based on a high-oil-price reference case scenario developed by the EIA and extended to 2050 by the committee. After starting at a few thousand HFCVs per year in 2012, the maximum practicable number of HFCVs on the road increases to almost 2 million in 2020, 60 million in 2035, and more than 200 million in 2050 for the Hydrogen Success case. As a percentage of all light-duty vehicles in service, this corresponds to approximately 0.7 percent, 18 percent, and 60 percent HCFVs in the U.S. fleet in these 3 years. The number of HFCVs grows rapidly after the market transition period, during which HFCVs are assumed to be supported by government subsidies and growing consumer acceptance. In the Hydrogen Success scenario, HFCVs compete only with improving conventional gasoline-powered vehicles and become cost-competitive by 2023, as discussed below in this summary.

CONCLUSION 5: In the judgment of the committee, the maximum practicable number of HFCVs that could be on the road by 2020 is around 2 million. Subsequently, this number could grow rapidly to as many as 60 million by 2035 and more than 200 million by midcentury, but such rapid and widespread deployment will require continued technical success, cost reductions from volume production, and government policies to sustain the introduction of HFCVs into the market during the transition period needed for technical progress. See Chapter 6.

Impact of Fuel Cell Vehicle Deployment on Reductions in Oil Use and CO_2 Emissions

Another objective of this study was to assess the potential of HFCVs to achieve significant reductions in oil imports and CO_2 emissions. Estimating such impacts is difficult because of the complexities and unknowns inherent in any analysis of future transportation systems and fuel options. For example, in recent years, energy prices and equipment costs have escalated dramatically; how these will vary in the future is unknown. The committee used its scenarios to inform its judgments about the potential magnitude and timing of reduced oil use and CO_2 emissions associated with adoption of HFCVs.

Figure S.2 shows results for the Hydrogen Success case relative to the reference case without fuel cell vehicles. As noted before, the reference case is based on an 2008 EIA high-oil-price scenario extended to 2050 by the committee. Gasoline use—taken as a measure of oil consumption and imports—is reduced by only about 1 percent by 2020; however, by 2035 the reduction in gasoline use grows to about 24 percent and by 2050 to nearly 70 percent. Similar trends are shown in Figure S.2 for potential reductions in well-to-wheels CO_2-equivalent emissions that account for all greenhouse gas emissions in the fuel supply chain (see Chapter 6). In the Hydrogen Success case, net annual greenhouse gas emissions are reduced by less than 1 percent in 2020, but subsequent reductions are much greater (i.e., a 20 percent reduction in 2035 and a reduction of more than 60 percent in 2050 compared to the reference case). These numerical estimates, especially for the longer term, are highly uncertain and sensitive to assumptions about the reference case as well as the Hydrogen Success scenario. Unlike savings in oil use, however, net reductions in CO_2 emissions do not automatically accrue from the use of hydrogen-fueled vehicles.

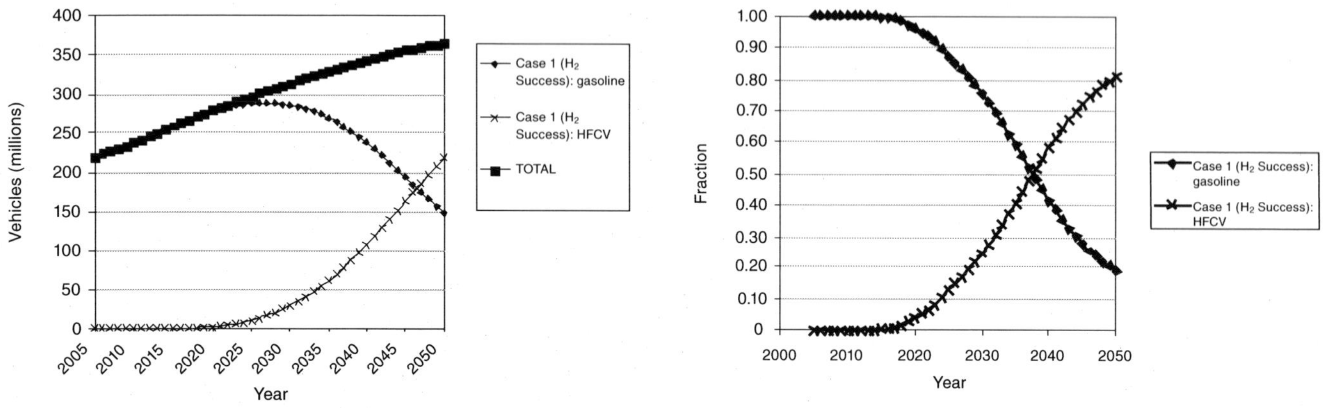

FIGURE S.1 (Left) Hydrogen fuel cell vehicles in the U.S. light-duty fleet and (right) fraction of new hydrogen vehicles sold each year for the Hydrogen Success case. This case assumes HFCVs compete only with gradually improving conventional gasoline-powered vehicles, and represents the committee's best estimate of the maximum practicable number of HFCVs deployable by 2020.

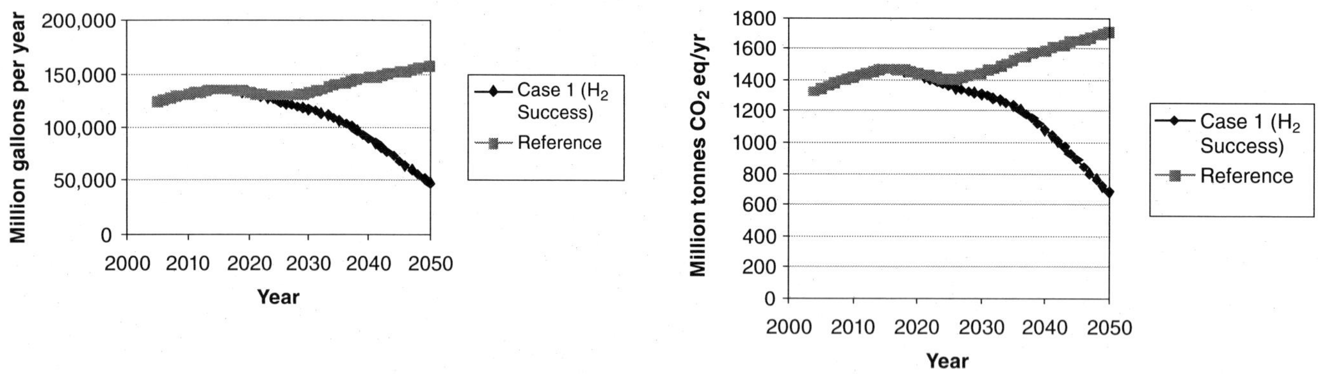

FIGURE S.2 (Left) Annual gasoline consumption and (right) annual well-to-wheels greenhouse gas emissions for the Hydrogen Success case relative to a reference case with no hydrogen vehicles. Case 1 assumes that HFCVs compete only with gradually improving conventional gasoline-powered vehicles.

Rather, the Hydrogen Success scenario assumes that policy measures are enacted prior to 2020 to incentivize or require the control of CO_2 emissions from the central stations used to produce hydrogen and that production from such plants begins around 2025, with hydrogen delivered by pipeline to refueling stations. Prior to that time, the production of hydrogen from distributed natural gas reformers results in CO_2 emissions, although at half the level of today's gasoline vehicles on a well-to-wheels basis.

CONCLUSION 6: While it will take several decades for HFCVs to have major impact, under the maximum practicable scenario fuel cell vehicles would lead to significant reductions in oil consumption and also significant reductions in CO_2 emissions if national policies are enacted to restrict CO_2 emissions from central hydrogen production plants. See Chapter 6.

Timetable for Market Transition

The potential benefits of reduced oil consumption and CO_2 emissions described above assume that HFCVs are deployed in increasing numbers according to the committee's Hydrogen Success scenario. Since HFCVs initially are far more expensive than conventional vehicles, the financial subsidy required to deploy them (and thus achieve future benefits) depends strongly on how long it takes HFCVs to compete economically in the marketplace with conventional gasoline vehicles. To estimate that transition period, the committee first estimated the total annual expenditures needed to purchase and operate increasing numbers of HFCVs as shown in Figure S.1. The unit cost of fuel cell vehicles was assumed to decline along a learning curve with increasing production. Hydrogen supply costs also declined with increasing production. These costs included the cost of energy feedstocks and other operating costs, plus the capital cost of infrastructure for refueling stations and hydrogen production facilities (including pipelines from central production plants in later years). Infrastructure cost estimates assumed the initial introduction of HFCVs in selected large cities and then in other locations over time, according to scenarios developed by DOE (see Chapter 6).

The committee then compared the year-by-year costs of purchasing and fueling the number of HFCVs dictated by the Hydrogen Success scenario to the annual cost of purchasing and fueling the same-size fleet of gasoline-powered vehicles. This analysis assumed that consumers would value fuel on a cost-per-mile-traveled basis rather than cost-per-gallon-equivalent. *When the sum of vehicle costs plus fuel costs for the HFCVs became less than that for gasoline vehicles, the hydrogen vehicles were projected to be economically competitive on a life-cycle basis*. That crossover was taken as the end of the transition period. For the Hydrogen Success scenario, that year was 2023. By that time a total of about 5.5 million fuel cell vehicles are produced, according to this scenario.

The breakeven year for competitiveness is sensitive to various assumptions, including HFCV costs and the costs of hydrogen and gasoline over time. However, the results described above are not significantly affected by recent changes in fuel economy standards for new vehicles.[2] Chapter 6 presents further details of the transition period analysis.

CONCLUSION 7: The unit costs of fuel cell vehicles and hydrogen in the Hydrogen Success scenario—the maximum practicable case—decline rapidly with increasing vehicle production, and by 2023 the cost premium for HFCVs relative to conventional gasoline vehicles is projected to be fully offset by the savings in fuel cost over the life of the vehicle relative to a reference case based on the EIA high-oil-price scenario. At that point, according to the committee's analysis, HFCVs become economically competitive in the marketplace. See Chapter 6.

A Budget Roadmap for the Market Transition

Producing and deploying the number of HFCVs shown in Figure S.1 will not happen by itself—significant government support will be needed to achieve this result. Even with such support there is no guarantee that the technical and economic performance improvements assumed in the Hydrogen Success case can be achieved in the given time frame. There are risks as well as potential rewards with these new technologies. Nonetheless, if the substantial potential benefits of hydrogen vehicles suggested in Figure S.2 are to be realized, substantial financial investments also will be required from both industry and government. As requested in its statement of task, the committee estimated a budget roadmap of total annual costs to government and industry for (1) conducting the research, development, and demonstration required for the transition to hydrogen and (2) deploying the maximum practicable number of fuel cell vehicles required for the transition. The estimated costs for these activities are summarized below.

Research, Development, and Demonstration Costs for the Transition

Because most future spending plans for RD&D are proprietary and there is very little information on which to base such estimates, those provided here are rough at best. The committee's estimates for government spending were based on budgets for DOE, the main government agency supporting RD&D on hydrogen. The DOE 2007 (budgeted) R&D funding and 2008 (requested) RD&D funding for the hydrogen program are each approximately $300 million—a level the committee judged to be adequate. Estimates of future budgets assumed that funding for individual programs was discontinued once a program was completed (such as the current program for distributed natural gas reforming), and that funds were added for projects that appeared appropriate for increases (such as biomass gasification). Funding for some areas was held constant (in constant dollars) when it was judged that considerable work was still needed (such as for fuel cells). These rough estimates of annual government RD&D funding were projected to 2023, the breakeven year for HFCVs in the Hydrogen Success case (see Chapter 7). The total RD&D funding requirement for 2008-2023 was estimated to be roughly $5 billion in constant 2005 dollars. This number could be adjusted up or down, depending on the need for new programs and demonstrations.

For the U.S. private sector, a current RD&D funding level of about $700 million per year was estimated based on a study commissioned in 2005 by a collaboration of several world fuel cell councils, adjusted for the United States (as discussed in Chapter 7) and supplemented by additional public information from small private U.S. companies. Much of the estimated spending needed in 2008 to 2012 would be for demonstration vehicles prior to commercial-scale manufacturing. Subsequently, private RD&D in conjunction with government programs would likely address remaining technical issues and opportunities for improvements to vehicle and hydrogen production technologies. Thus, total U.S. private spending on RD&D for the hydrogen transition from 2008 to 2023 was estimated to continue at an average of $700 million per year, totaling roughly $11 billion in constant 2005 dollars.

Although no RD&D funding estimates were projected beyond 2023, the committee fully anticipates the need for continued RD&D funding by both government and the private sector in the Hydrogen Success scenario. In particular, increased emphasis on hydrogen production with low or

[2]The CAFE standards enacted in December 2007 phase in a 35 mpg requirement for new cars from 2011 to 2020.

zero carbon emissions—such as from renewable energy sources—is expected to be an area of growing importance, both during and after a transition to HFCVs.

CONCLUSION 8: The committee estimates that total government-industry spending on RD&D needed to facilitate the transition to HFCVs is roughly $16 billion over the 16-year period from 2008 through 2023, of which about 30 percent (roughly $5 billion) would come from U.S. government sources. Government and private spending beyond 2023 also will likely be required to support longer-term needs, but such estimates were beyond the scope of this study. See Chapter 7.

Vehicle Deployment Costs for the Transition

The committee's estimate of private plus government expenditures required to deploy the maximum practicable number of fuel cell vehicles by 2020 and beyond (as reflected by the Hydrogen Success scenario) is shown in Figure S.3, which indicates total annual costs for the two main components of commercial deployment: vehicle costs and hydrogen fuel costs. Capital investments for infrastructure to produce and deliver hydrogen account for approximately half of the annual hydrogen costs, with the remaining half being operating costs, mainly the cost of natural gas for hydrogen production.

Total annual costs grow rapidly as increasing numbers of vehicles are deployed. The cumulative cost of fuel cell vehicles purchased during the transition period 2012-2023 is about $170 billion, or an average of $30,000 per vehicle (see Table S.1). Of this, $40 billion (an average of $7,000 per vehicle) represents the additional investment for HFCVs over the roughly $130 billion "base vehicle" cost of an equivalent number of conventional vehicles. The additional cost of supplying hydrogen over this period is $16 billion. Thus, the cumulative expenditure for the transition totals $184 billion, most of which (91 percent) is for the production of vehicles, with the remaining 9 percent for hydrogen supply (roughly half for infrastructure and half for operating costs).

If the RD&D costs discussed above are added to the vehicle and hydrogen production costs, the budget roadmap shown in Figure S.3 would increase by approximately $1 billion per year ($16 billion total). This would bring the cumulative total to $200 billion for the period 2008-2023. Table S.1 summarizes the cumulative costs of the budget roadmap.

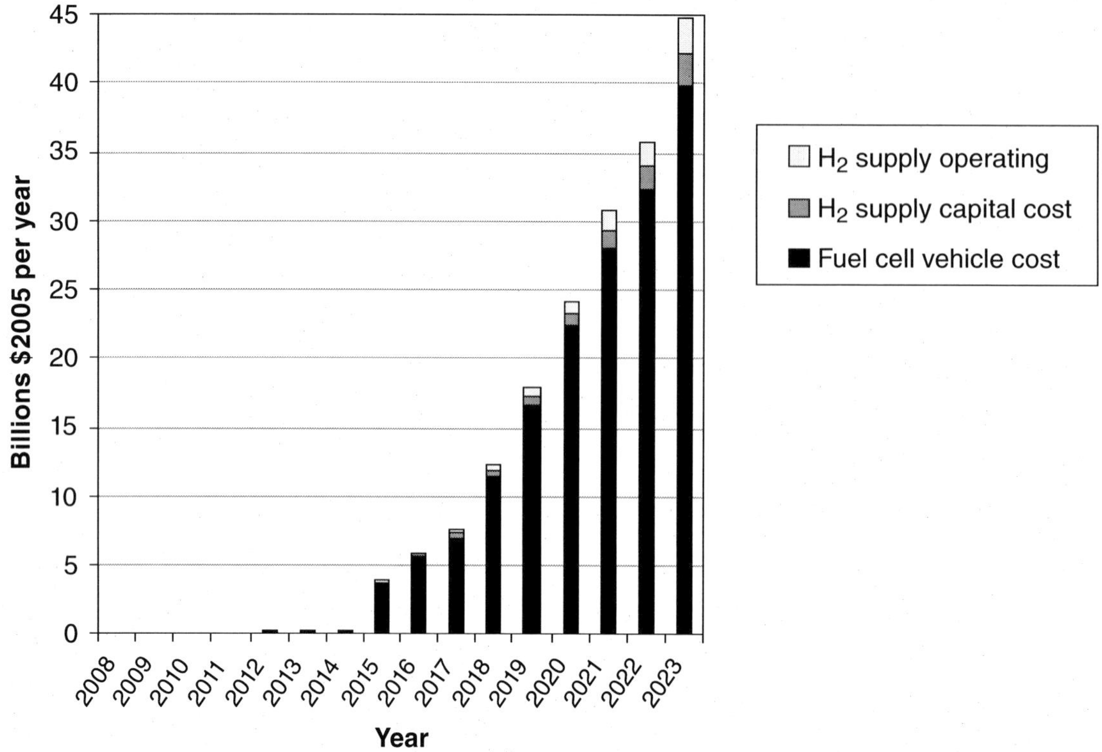

FIGURE S.3 Total annual expenditures for vehicles and hydrogen supply for transition to the breakeven year for the Hydrogen Success case, excluding RD&D costs. The cumulative cost, shared by government and industry, totals $184 billion, of which 91 percent is the cost of fuel cell vehicles and 9 percent is the cost of hydrogen supply (about half for infrastructure costs and half for additional operating costs, mainly natural gas feedstock). The additional $16 billion in private plus government RD&D costs over this period would bring the overall total to $200 billion.

TABLE S.1 Summary of Cumulative Budget Roadmap Costs for Transition to Hydrogen Fuel Cell Vehicles (maximum practicable number of vehicles by 2020)

Cost Elements	Total Cumulative Cost, 2008-2023	Average Cost per HFCV on Road 2008-2023[a]
"Base vehicle" cost of conventional vehicles	$128 billion	$23,000
Average incremental fuel cell vehicle cost relative to conventional gasoline vehicles	$40 billion	$7,000[b]
Total purchase cost of fuel cell vehicles	$168 billion	$30,000[c]
Infrastructure capital cost for hydrogen supply	$8 billion	$1,500
Total operating cost for hydrogen supply	$8 billion	$1,500
Total cost of hydrogen supply	$16 billion	$3,000
Total cost for vehicles and hydrogen fuel supply	$184 billion	$33,000
Estimated government share of total vehicle and hydrogen fuel supply cost	$50 billion	$8,500
Government RD&D funding	$5 billion	$1,000
Private RD&D funding	$11 billion	$2,000
Total funding for government and private RD&D	$16 billion	$3,000
Total cost for vehicles, hydrogen, and all RD&D	$200 billion[d]	$36,000
Estimated government share of total cost for vehicles, hydrogen, and RD&D	$55 billion	$9,500

[a]Rounded estimates based on 5.54 million HFCVs on the road in 2023.
[b]The final (learned-out) incremental cost per vehicle in 2023 is $3,600.
[c]The final (learned-out) cost per vehicle in 2023 is $27,000.
[d]Includes $128 billion "base vehicle" cost of conventional vehicles that would have been purchased instead of HFCVs.

NOTE: All costs in constant 2005 U.S. dollars.

Beyond 2023 the hydrogen vehicle and infrastructure system pays for itself.

Government Share of Transition Costs

The question of how the vehicle production and hydrogen supply costs shown in Figure S.3 should be shared between the federal government and private industry has no simple or single answer. In the committee's judgment, a realistic estimate of the government share of total costs to facilitate the maximum practicable transition to HFCVs (based on the Hydrogen Success scenario) would be the incremental cost of purchasing HFCVs, plus about half the total cost of building and operating the infrastructure needed to supply hydrogen during the transition period. Those incremental costs are shown in Figure S.4. In this case, the added cost for vehicles totals $40 billion over the transition period (as noted above), while hydrogen infrastructure costs add another $8 billion. Various factors could either reduce or raise these costs to some degree (see Chapter 7). The committee estimated the total government cost for this scenario to be approximately $50 billion (an average of $8,500 per vehicle during the transition period).

Adding government RD&D costs to the figures above would increase the budget roadmap by approximately $300 million per year, or a total of $5 billion. This would bring the cumulative government investment to $5 billion (an average of $9,500 per vehicle) from 2008 through 2023, as summarized in Table S.1. This translates to an average of roughly $3 billion per year over 16 years (2008-2023). To put these amounts in perspective, the U.S. government subsidy for ethanol fuel in 2006 was approximately $2.5 billion and, if extended at the current rate, could grow to $15 billion per year in 2020 due to the recently enacted (December 2007) energy bill.[3]

[3]The Volumetric Ethanol Excise Tax Credit (VEETC) of 51 cents per gallon benefits all ethanol blended with gasoline, which was about 5 billion gallons in 2006, according to DOE data. Although the VEETC is set to expire after 2010, Congress is debating various ways of extending it, as it has since the credit was first created in 1978. The Energy Independence and Security Act of 2007 established a renewable fuel standard that would reach 30 billion gallons by 2020, most of which is likely to be ethanol. A credit of 51 cents per gallon applied to that amount would represent a subsidy in excess of $15 billion per year.

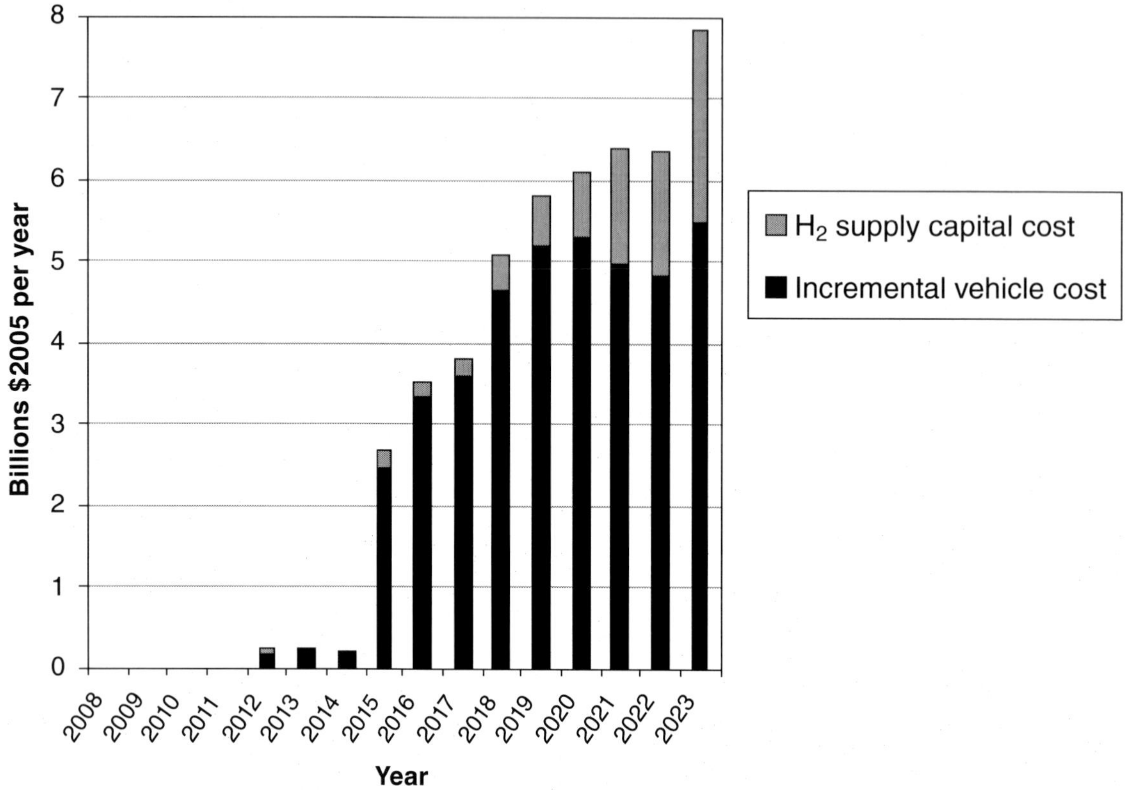

FIGURE S.4 Annual government expenditures through the transition to 2023. Estimated expenditures are based only on the incremental costs of fuel cell vehicles over conventional vehicles, plus the capital cost for hydrogen infrastructure, for the Hydrogen Success scenario (excluding RD&D costs). The cumulative cost is $48 billion, of which 83 percent is the cost of vehicles and 16 percent is the cost of hydrogen infrastructure. Government RD&D costs over this period total an additional $5 billion.

Other Cost Considerations

While the committee's budget roadmap considered only the funding required to launch the maximum practicable scenario for fuel cell vehicles, it is unlikely that federal funding would terminate after the breakeven year for transition, as assumed here. Rather, continued expenditures at some level would be expected, although the committee did not attempt to estimate such requirements.

Note, too, that the budget roadmaps presented here do not reflect the savings to consumers from reduced expenditures for gasoline during 2012-2023 (estimated at roughly $17 billion) or the loss of government tax revenues from gasoline sales displaced by hydrogen (roughly $5 billion). Also, the budget estimates do not include any costs for technical educational or training programs to support the transition, because the committee estimated no shortage of workers with the needed skills during this period (see Chapter 7). Other types of training programs (e.g., safety training) are likely to be needed.

There is considerable uncertainty in predicting the costs of deploying HFCVs, in particular technical success, oil prices, and carbon policy. To the extent that HFCVs exceed the technical and cost targets assumed in the maximum practicable analysis, the government's share of costs could be reduced. Similarly, to the extent that HFCV imports from non-U.S. automakers contribute to the hydrogen transition, the magnitude of U.S. government-supported vehicle costs also would be reduced. But progress may be slower than assumed here, and pushing HFCVs into the market would then be more expensive than shown in Figure S.3. Before companies and the government start ramping up the funding for the transition in about 2015, it will be important to fully assess the state of the technology and expectations for the market. Insofar as progress is either faster or slower than expected, it will be important to adjust policies in response, while avoiding the perception that promotional policies are not durable.

Finally, the committee notes that the budget roadmaps developed in this study apply only to the transition period through 2023. However, the successful introduction of HFCVs also would involve substantial longer-term expenditures—primarily by the private sector—for infrastructure, energy resources, and other requirements of a full-scale HFCV-based transportation system. Estimates of longer-term capital requirements and other resource needs for the Hydrogen Success scenario can be found in Chapters 6 and 7 of this report, along with a discussion of issues such as long-term labor force requirements that remain for future study.

CONCLUSION 9: The estimated government cost to support a transition to hydrogen fuel cell vehicles is roughly \$55 billion over the 16-year period from 2008 to 2023, primarily for the production of fuel cell vehicles (\$40 billion of incremental cost) and, to a lesser extent, for the initial deployment of hydrogen supply infrastructure (about \$10 billion) and R&D (about \$5 billion). No shortages are foreseen in the critical workforce skills needed to accomplish the transition. However, further study is necessary to assess the longer-term costs, institutional issues, workforce issues, and impacts of undertaking the major hydrogen infrastructure development required to support widespread use of HFCVs. See Chapter 7.

Actions Required by Government to Implement the Transition

Six types of actions and policies are needed to exploit the potential of hydrogen to reduce oil use and CO_2 emissions from the transportation sector, consistent with the Hydrogen Success case described in this report:

1. Actions to ensure continued development of HFCVs technologies,
2. Actions to deal with the high initial cost of HFCVs,
3. Actions to develop the initial infrastructure needed to support HFCVs,
4. Actions to reduce the cost of the initial distributed infrastructure for hydrogen,
5. Policies that promote energy security by improving the end use efficiency of transportation fuels, and
6. Policies that limit greenhouse gas emissions to ensure that hydrogen produced using domestic energy resources results in little or no emission of CO_2 to the atmosphere.

More specifically, to stimulate RD&D to overcome remaining technical hurdles and encourage long-run penetration of hydrogen vehicles, a technology-push approach (i.e., not relying simply on market forces) is required. Such an approach would have to employ policies and incentives that are carefully targeted, substantial, durable, and gradually phased out over time with continued technology progress. However, the design and the choice of such policies are neither simple nor straightforward.

Targeted policies to kick-start a market for hydrogen vehicles could include such measures as federal tax credits, subsidies for hydrogen vehicle purchases, or minimum sales share quotas imposed on vehicle manufacturers. If financial incentives are used, they would have to be substantial, given that life-cycle costs for hydrogen vehicles are currently much higher than for comparable gasoline vehicles. Without large incentives, consumers also may be reluctant to switch in significant numbers to a new and unfamiliar type of vehicle and fuel. As noted above, policy measures that significantly limit CO_2 emissions also will be required to ensure that hydrogen is produced in ways that do not add to the burden of greenhouse gas emissions.

Durable incentives, lasting 15-20 years or more, also would be critical for setting private sector expectations about the long-run payoffs to investments with high up-front costs. However, subsidies should be progressively phased out over time as long-term penetration targets are approached, in order to limit government funding requirements and encourage firms to act more quickly in the earlier years of the program. Coordination of financial incentives with the technical progress of the program is therefore crucial.

Ultimately, however, the heart of any policy actions to promote substantial HFCV penetration must be the incentives or requirements for auto manufacturers to develop and mass-produce hydrogen vehicles consistent with the budget roadmaps presented above and motivated by the goals of national energy and environmental policies. Chapter 8 provides further discussions of policy design.

CONCLUSION 10: Policies designed to accelerate the penetration of HFCVs into the U.S. vehicle market will be required to exploit the long-term potential of HFCVs. The committee concluded that these policies must be durable over the transition time frame but should be structured so that they are tied to technology and market progress, with any subsidies phased out over time. Such policies are likely to deliver significant long-term reductions in U.S. oil demand, but additional policies limiting greenhouse gas emissions will be required in order to also reduce CO_2 emissions significantly. See Chapter 8.

SYNERGIES WITH THE ELECTRIC POWER SECTOR

The committee also was asked to consider the role that hydrogen in stationary electric power might play in stimulating a transition to hydrogen-fueled vehicles. There are potential synergies between the transportation and electric power sectors that could benefit both sectors in the near term and longer term.

In the near term, the electric power sector has the potential to accelerate the volume of hydrogen available for fuel cell vehicles during the transition period by producing electricity targeted for electrolysis of water to generate hydrogen for transportation. While large central electrolyzers appear not to be competitive with other central plant technologies, the use of small-scale electrolyzers at a refueling site could play an important role during the start-up phase (from about 2012 to 2025), when the cost burden of larger-scale natural gas reforming plants is a potential barrier to hydrogen supply. The committee's analysis indicated that small-scale electrolysis, although in general more expensive than hydrogen from distributed natural gas reforming, might nonetheless supply hydrogen to areas with limited access to natural gas or to other locations in particular situations. Electric utilities, working with state utility commissions, could provide economic incentives to reduce the cost burden of the electrolysis process. The impact of electricity cost on hydrogen production cost is discussed in Chapter 3. In addition, DOE and several companies have R&D programs to reduce the cost of electrolysis.

In the longer term (e.g., 2025 through 2050), there is a potential for synergy in the use of hydrogen for stationary power generation and for transportation. The key enabling technologies envisioned as sources of hydrogen in the committee's scenarios are gasification plants fed either by biomass or by coal with CCS. Coal-based hydrogen at the scale envisioned is anticipated to use the same type of equipment used in integrated gasification combined cycle (IGCC) power plants projected to be built in the future. Thus, hydrogen produced from gasification can be used as a fuel either for stationary power generation or for export to the transportation market. For coal-based hydrogen production, CCS will be required to avoid significant CO_2 emissions. Biomass gasification, however, is a potentially carbon-neutral source of hydrogen. Although other power plant technologies, such as nuclear and renewable energy systems, also could be used to generate carbon-free hydrogen (mainly via electrolysis), gasification-based systems are the most economical approach in the committee's assessment.

A potential synergy not explicitly modeled in the committee's analysis is that IGCC plants with CCS also offer a potential remedy to the "chicken-and-egg" problem of providing incentives for initial investments in large-scale hydrogen production facilities needed to reduce future hydrogen costs during and after the transition period. The flexibility of gasification systems to provide electric power as well as hydrogen can significantly reduce the financial risks associated with large-scale hydrogen production during the scale-up phase of HFCV commercialization.

As noted above in the discussion of technical readiness, several key issues still must be resolved to achieve the potential synergies described. Foremost is the need to demonstrate the production of coal-based hydrogen or electricity with CCS at a commercial scale (e.g., an electrical equivalent of several hundred megawatts) within the next decade. In the committee's view, such demonstrations are required to enable decisions about the applicability and deployment of central hydrogen production with CCS in the 2025 time frame, as assumed in the Hydrogen Success scenario. Again, utilities and other companies, working with their regulatory commissions, could be given incentives to pursue the large-scale demonstration of these technologies more rapidly.

CONCLUSION 11: With appropriate policies or market conditions in place, potential synergies between the transportation sector and the electric power sector could accelerate the potential for reduced oil use and decreased CO_2 emissions as benefits from the use of hydrogen in both sectors. In the near term, electrolysis of water at refueling sites using off-peak power, and in the longer term (after 2025), cogeneration of low-carbon hydrogen and electricity in gasification-based energy plants, are potential options that offer additional synergies. See Chapter 5.

POTENTIAL OF ALTERNATIVE TECHNOLOGIES

The committee also was asked to consider whether other technologies would be less expensive or could be implemented more quickly than HFCVs to achieve significant reductions in CO_2 emissions and oil imports. The committee concluded that a rigorous assessment of the costs and impacts of all technologies that compete with HFCVs would require a level of effort beyond the scope and resources of the current study. Thus, the committee approached this task as one intended to provide context for the development of HFCVs—not one intended to identify specific "technology winners" that might be preferable to HFCVs. For purposes of analysis, the committee therefore chose to focus on the impacts, but not the costs, of one vehicle alternative and one fuel alternative for reducing oil use and CO_2 emissions: (1) evolutionary internal combustion engine vehicles (ICEVs), including gasoline-hybrid electric vehicles, and (2) the use of biofuels—specifically ethanol and biodiesel—to replace petroleum-based fuels. The committee did not evaluate the impact of other vehicle or fuel technologies as explained above, but it recognizes the potential of advanced technologies such as PHEVs to achieve significant reductions in oil use and CO_2 emissions if they are successfully deployed. Such technologies are discussed further in Chapters 4-6.

The following sections briefly summarize the committee's assumptions regarding potential developments in advanced conventional vehicles and biofuels. To assess the potential impacts of such developments on reducing oil use and CO_2 emissions, the committee analyzed these assumptions using the same analytical model it employed to assess potential impacts of HFCVs. In addition, the committee attempted to match the degree of technological optimism and aggressive implementation of the Hydrogen Success case for both alternative cases. Results of those analyses also are summarized below.

Advanced Conventional Vehicles

Conventional power trains and vehicles have continued to improve since the invention of the automobile more than 100 years ago. The committee found significant potential to continue this evolution through improvements in engine and transmission efficiency, aerodynamic design, and reductions in rolling resistance, weight, and accessory loads. Much of the recent progress in these areas has been directed to enhancing vehicle performance and size rather than reducing fuel consumption. However, if future improvements to gasoline-powered vehicles were used exclusively for fuel economy gains, this study estimates that oil consumption and greenhouse gas emissions per mile of travel for new vehicles could be reduced by almost 30 percent in 2020, more than 40 percent by 2035, and about 50 percent by 2050 compared to average vehicles in 2006. In all cases, however, policy measures (such as the 2007 CAFE standards) and/or significant long-term increases in fuel costs probably will be required to realize these potential fuel economy gains in a significant number of on-road vehicles. Absent such policies or large fuel cost increases, the committee expects that much of the conventional technology potential will either remain unused or be directed to attributes other than fuel economy.

The committee further estimated that evolutionary improvements in current gasoline-electric hybrid vehicles—a more revolutionary and more recent approach to power train design—could reduce fuel consumption and greenhouse gas emissions per mile for new vehicles by about 50 percent in 2020, more than 60 percent by 2035, and nearly 70 percent by 2050 compared to today's conventional gasoline vehicles. Reaching this full potential will likely require adoption of many of the evolutionary vehicle and power train improvements for conventional vehicles, as well as further progress on battery technology. Chapter 4 elaborates on the technical basis for the efficiency improvements outlined above. These technologies will increase the initial costs of vehicles, but savings in fuel cost will accrue over the life of the vehicle and thus could offset costs to the consumer on a life-cycle basis.

CONCLUSION 12: Continued advancements in conventional vehicles offer significant potential to reduce oil use and CO_2 emissions through improved fuel economy, but policy measures and/or significant long-term increases in fuel cost probably will be required to realize these potential fuel economy gains in a significant number of on-road vehicles. See Chapter 4.

Biofuels

Automotive fuels produced from crops or other forms of biomass have the potential to further reduce oil imports and CO_2 emissions. As in the case of vehicle technologies, biomass feedstocks and conversion technologies span a range of levels of development and maturity. There is also a corresponding lack of information about when processes in the early stages of development might become commercial. Therefore, this study used options for which more data were available to assess the potential of biofuels to offer earlier reductions in oil use and CO_2 emissions compared to HFCVs.

Such assessments are by no means straightforward because of the complexities and uncertainties in accounting for oil use and greenhouse gas emissions across the chain of processes involved in biofuel production (see Chapter 4). Recent papers in the literature addressing the potential for soil root carbon CO_2 release from land use changes indicate the difficulty of these complexities. The most common biofuel produced in the United States is ethanol made from fermenting corn kernels. Because the energy content of ethanol is less than that of gasoline, roughly 30 percent more ethanol by volume is needed to replace each gallon of gasoline. Furthermore, because petroleum-based fuels are used in growing and producing corn ethanol, net reductions in both oil use and greenhouse gas emissions are greatly diminished relative to the impacts of equivalent energy from gasoline. Although some agricultural and ethanol production practices could yield greater reductions (such as ethanol produced from sugar cane), others—such as processes using coal-based energy or the clearing of existing forest to plant corn—could increase greenhouse gas emissions compared to gasoline. The committee also found that the potential to drive up food prices made it unlikely that more than about 25 percent of U.S. corn crops would be devoted to ethanol, which would limit corn-based ethanol to about 12 billion gallons after 2015 (the energy equivalent of less than 6 percent of the reference case gasoline demand for light-duty vehicles in that year and a still smaller percentage in later years).

The technology to produce cellulosic ethanol—ethanol from woody biomass, grasses, or crop residues—is not yet demonstrated for commercial production but is actively being developed. The committee found that if successful, it could reduce oil use by more than 90 percent and greenhouse gas emissions by roughly 85 percent compared to equivalent gasoline use in light-duty vehicles. The committee estimated that domestic resources for cellulosic biomass would allow production of about 45 billion to 60 billion gallons of cellulosic ethanol by 2050, potentially displacing about 20 percent of the baseline gasoline demand estimate for that year.

Biodiesel, a fuel produced from animal or plant oils such as soy, would reduce oil use and greenhouse gas emissions similarly to cellulosic ethanol. Producing biodiesel is a simpler process that is already commercial. If biodiesel were produced from current sources (e.g., soybeans), the potential impact on commodity prices would limit its potential to roughly 30 percent of the soy crop, which would provide slightly more than 1 billion gallons of biodiesel by 2020. This would displace a very small fraction of diesel demand in that year, based on committee estimates. The potential for biodiesel fuels was thus judged to be small unless breakthroughs are achieved in areas such as oil production from algae.

CONCLUSION 13: Although use of corn- and oil-based biofuels can provide some benefits in reducing U.S. oil use and CO_2 emissions, cellulosic biofuels will be required

for such benefits to be significant. Lower-cost biofuel production methods and conversion processes will have to be developed for large-scale commercialization, but the initial high costs of biofuels, together with other barriers, may limit their market potential, absent policy interventions or significant oil price increases or supply disruptions. See Chapter 4.

Impacts of Alternative Technologies on Oil Use and CO_2 Emissions

To evaluate whether alternative technologies might be implemented more quickly than HFCVs to achieve significant reductions in oil use and CO_2 emissions, the committee extended the modeling framework described above for HFCVs to include two alternatives—evolutionary vehicles and biofuel technologies. As with hydrogen, such modeling estimates are uncertain because of the complexities and unknowns inherent in any analysis of future transportation systems and fuel options. Insights from modeling were nonetheless of significant value in informing the committee's judgment about the potential impacts of alternatives to hydrogen.

Toward this end, the committee developed and analyzed additional scenarios for the two selected alternative technologies with technological optimism and aggressive implementation similar to those for the Hydrogen Success case (the maximum practicable case). Case 2 (ICEV Efficiency) focused on improvements to conventional vehicles. This case shows (based on the analysis in Chapter 4) that aggressive implementation of evolutionary technology improvements for gasoline vehicles raised the average on-road fuel economy (which is typically 20 percent lower than the Environmental Protection Agency's "sticker" miles per gallon [mpg]) of the light-duty fleet to about 30 mpg by 2020 and to nearly 40 mpg by 2035, with a small additional improvement by 2050. Conventional hybrid vehicles were estimated to improve to about 45 mpg by 2020, and then to about 55 mpg by 2035 and about 60 mpg by 2050. In Case 2, the growing penetration of hybrids gained them an 80 percent share of the total vehicle market by 2050 (see Chapter 6). These estimates assume that the evolutionary technologies result in efficiency improvements and that consumers buy them.

Case 3 (Biofuels) assumed aggressive development and use of biofuels to power the conventional vehicles of the baseline scenario. Most of this biofuel was in the form of cellulosic ethanol, which was to reach commercialization by 2010 (based on DOE's biofuels roadmap), followed by rapid expansion to 16 billion gallons per year in 2020, 32 billion gallons per year in 2035, and 63 billion gallons per year in 2050. Grain-based ethanol production was assumed to reach a maximum of 12 billion gallons per year by 2015 and to remain at that level through 2050.

Full details of the assumptions used to analyze alternative vehicles and biofuels appear in Chapters 4 and 6 of this report. Here, the major results and implications of that analysis are highlighted.

The results from Case 2 (ICEV Efficiency) (Figure S.5) indicate that aggressive fuel economy improvements in conventional light-duty and hybrid vehicles follow the reference case, which includes the 2007 CAFE standards through 2020, but then could potentially deliver greater reductions in U.S. oil demand and CO_2 emissions compared to the Hydrogen Success scenario, through about 2040. Subsequently, under the assumptions of this scenario, the rates of growth in the benefits of potential efficiency improvements begin to slow at a time when benefits from the Hydrogen Success case are still increasing. Breakthroughs and rapid market penetration in other developing vehicle technologies, such as plug-in hybrids and diesel hybrids, potentially could increase the benefits of reduced oil consumption and CO_2 emissions above those shown in Case 2 (ICEV Efficiency), particularly in the 2030-2050 time frame.

The results of Case 3 (Biofuels) (also shown in Figure S.5) suggest that biofuels alone also could potentially reduce oil

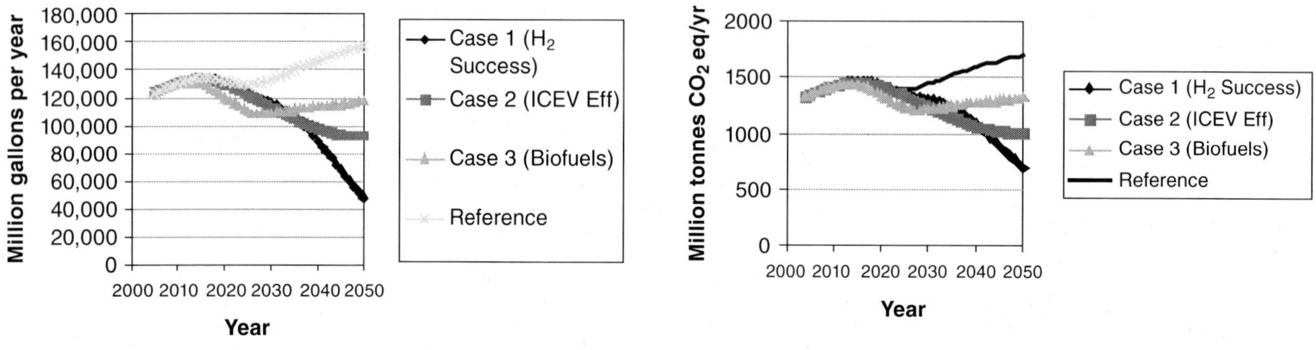

FIGURE S.5 Comparison of (left) annual gasoline use and (right) annual greenhouse gas emissions (as equivalent CO_2) for Cases 1-3 compared with the reference case.

demand and greenhouse gas emissions (measured as equivalent CO_2 of all greenhouse gas emissions over the fuel cycle) sooner than the Hydrogen Success case, if cellulosic ethanol comes online by 2010 and grows aggressively thereafter. As with Case 2 (ICEV) efficiency improvements, however, the rate of growth in benefits from biofuels implementation begins to slow toward the end of the analysis period. In 2030-2040, the Hydrogen Success case has the potential to provide greater reductions and, by 2040, delivers two to three times the reductions in oil use and CO_2 emissions as the aggressive biofuels scenario.

Chapter 6 also presents the results of several additional cases involving combinations of alternative technologies. Although these scenarios give quantitative results different from those shown here, the qualitative conclusion is similar—that is, alternative technologies can deliver significant oil use and CO_2 emission reduction benefits earlier than HFCVs, but the largest sustained longer-term benefits are achieved using hydrogen fuel cell vehicles.

CONCLUSION 14: The committee's analysis indicates that at least two alternatives to HFCVs—advanced conventional vehicles and biofuels—have the potential to provide significant reductions in projected oil imports and CO_2 emissions. However, the rate of growth of benefits from each of these two measures slows after two or three decades, while the growth rate of projected benefits from fuel cell vehicles is still increasing. The deepest cuts in oil use and CO_2 emissions after about 2040 would come from hydrogen. See Chapter 6.

BENEFITS OF A PORTFOLIO APPROACH

Based on a comparison of the three scenarios in Figure S.5, the committee concluded that no single approach is likely to deliver both significant midterm and long-term reductions in oil demand and greenhouse gas emissions. Thus, conventional and hybrid vehicle technology, biofuels, and HFCVs should be considered not as competitors over the next few decades, but as part of a portfolio of options with a potential to deliver significant energy security and environmental benefits across a variety of time horizons. Other technologies not analyzed in this study, such as plug-in hybrids, battery electric vehicles, and other types of internal combustion engines, also should be examined as potential candidates for this portfolio. As in other domains, a portfolio of technology options is most likely to improve the chances of success while reducing the risks in the event that any one option fails to deliver on its promise.

Because advanced conventional vehicles, hybrid vehicles, and biofuels can deliver benefits in a shorter time frame, they may be able to more quickly reduce the potential impacts of climate change and reliance on oil imports, while also providing the time needed to further develop and commercialize hydrogen-based fuel cell technologies. Also, should the impacts of climate change or oil shocks mobilize an aggressive policy response, acceleration of HFCVs into the market could provide a path toward a zero-petroleum and potentially low-carbon option that can persist beyond the large, but eventually limited, potential of vehicle efficiency and biofuels alone.

To explore the value of a portfolio approach, the committee constructed Case 4 (Portfolio), which combines all three options of vehicle efficiency, biofuels, and HFCVs (see Chapter 6). Compared to the reference baseline scenario, the results showed that this portfolio of options has the potential to nearly eliminate oil demand from light-duty vehicles by the middle of the century, while reducing greenhouse gas emissions by almost a factor of 10 relative to the assumed baseline case (Figure S.6). Achieving this potential is likely to require a portfolio approach that takes advantage of the synergies among these technologies. For example, many of the technologies needed to improve the fuel economy of conventional vehicles, including weight reduction, improved aerodynamics, lower rolling resistance, and low accessories loads, also will be essential for fuel cell vehicles to reach the

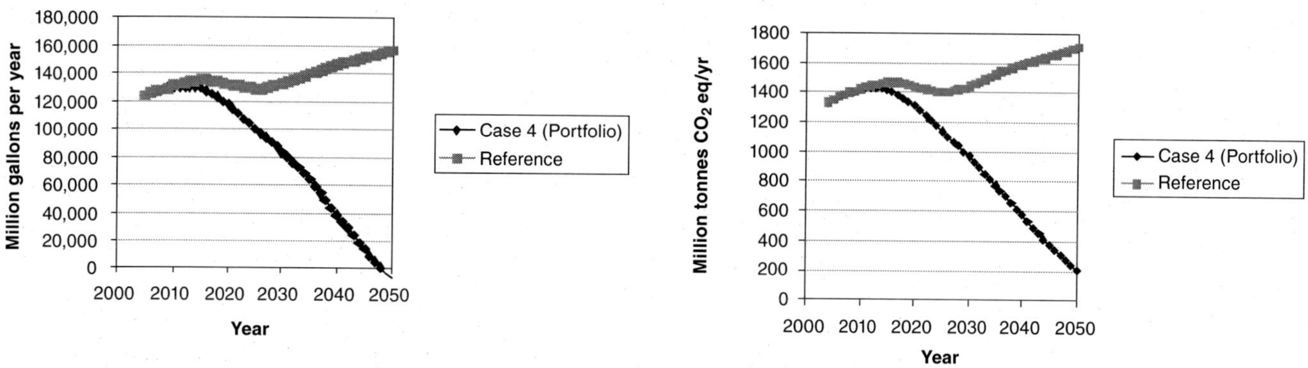

FIGURE S.6 Impact of combining the potential of HFCVs with advanced conventional vehicles, hybrid vehicles, and biofuels in a portfolio approach—Case 4: (Left) annual gasoline use and (right) annual greenhouse gas emissions.

efficiency levels assumed in this study. Similarly, advancements in battery and electronics technology for hybrids will likely find direct application in fuel cell vehicles, while some technologies used to produce biofuels also could be adopted for hydrogen production (such as gasification of biomass or on-site reformers using biomass).

CONCLUSION 15 (same as Conclusion 1): A portfolio of technologies including hydrogen fuel cell vehicles, improved efficiency of conventional vehicles, hybrids, and use of biofuels—in conjunction with required new policy drivers—has the potential to nearly eliminate gasoline use in light-duty vehicles by the middle of this century, while reducing fleet greenhouse gas emissions to less than 20 percent of current levels. This portfolio approach provides a hedge against potential shortfalls in any one technological approach and improves the probability that the United States can meet its energy and environmental goals. Other technologies also may hold promise as part of a portfolio, but further study is required to assess their potential impacts. See Chapter 9.

REFERENCES

EIA (Energy Information Administration). 2008. *Annual Energy Outlook 2008: With Projections to 2030*. Report DOE/EIA-0383. Washington, D.C.

Greene, D., P. Leiby, and D. Bowman. 2007. *Integrated Analysis of Market Transformation Scenarios with HyTrans*. Oak Ridge National Laboratory, Oak Ridge, Tenn.

Gronich, S. 2007. 2010-2025 Hydrogen Scenario Analysis. Presentation to the committee, February 20.

NRC (National Research Council). 2004. *The Hydrogen Economy: Opportunities, Costs, Barriers, and R&D Needs*. Washington, D.C.: The National Academies Press.

1

Introduction

Hydrogen fuel cell vehicles show great promise for alleviating two of the most difficult issues facing the nation: climate change and the insecurity of oil imports. Fuel cells convert the chemical energy of hydrogen directly to electricity, which drives the vehicle. The process is very efficient and clean, especially relative to internal combustion engines. The only emission is water, the result of combining hydrogen with oxygen from the atmosphere. No gasoline is used, and no carbon dioxide (CO_2), the main greenhouse gas that causes climate change, is produced or emitted from the vehicle.

That promise, however, will be realized only after some very substantial challenges are met:

- Hydrogen fuel cell vehicles (HFCVs) currently are extremely expensive, several hundred thousand dollars each, in part because the few that are operating are hand built, with no benefits of mass production. However, the fuel cells themselves and on-board storage of hydrogen are very expensive. Considerable research and development (R&D) are necessary to bring these costs down to where HFCVs will be competitive with conventional vehicles.
- Hydrogen is not a natural resource but must be manufactured from a primary energy source (e.g., natural gas or coal) or from electrolysis of water, which requires electricity. More energy is required to produce hydrogen than can be recovered from the hydrogen. All current methods of producing commercial hydrogen also emit significant amounts of CO_2 per unit of hydrogen produced, either from the feedstock fuel or from the fuels used to generate electricity. Although not now prohibitively expensive, hydrogen costs also must be reduced by R&D, especially if it is to be produced with little or no emission of carbon dioxide.
- There is currently no infrastructure for supplying hydrogen to HFCVs. The very few hydrogen filling stations in the country were built to service test vehicles. Thousands will have to be built if hydrogen is to become the dominant transportation fuel. In addition, facilities for producing the hydrogen and delivering it to the filling stations must be built.
- Even though projections show that hydrogen HFCVs could be the vehicle of choice in several decades, the transition to hydrogen fuel cell vehicles will be challenging. Sufficient filling stations must be available to assure early purchasers of HFCVs that they will be able to operate their vehicles, but companies will be reluctant to build the stations until there is business for them.

These issues, especially the potential for reductions in the use of oil and emissions of carbon dioxide, prompted the U.S. Congress to request this study in the Energy Policy Act of 2005. In response, the U.S. Department of Energy (DOE) arranged for this study from the National Research Council (NRC). It focuses on the resources that will be needed to bring fuel cell and hydrogen technologies to readiness and then get them through the transition to the point that they will be self-supporting. The NRC established the Committee on Assessment of Resource Needs for Fuel Cell and Hydrogen Technologies, which was guided by the following statement of task:

> The National Academy of Sciences National Research Council will appoint a committee to carry out a study of fuel cell technologies that provides a budget roadmap (e.g., what investments in R&D, demonstrations, skilled people, infrastructure will be required) for the development of fuel cell technologies and the transition from petroleum to hydrogen in a significant percentage of the vehicles sold by 2020. The committee will:
>
> (1) establish as a goal the maximum percentage practicable of vehicles that the committee determines can be fueled by hydrogen by 2020;
> (2) determine the amount of funding required as a whole (public and private), and to the extent possible the Federal investments required, to meet the goal established under paragraph (1). Consideration will be given to investments

needed for R&D, demonstrations, skilled people, and infrastructure;
(3) determine what actions are required to meet the goal established under paragraph (1);
(4) examine the need for expanded and enhanced Federal research and development programs, changes in regulations, grant programs, partnerships between the Federal Government and industry, private sector investments, infrastructure investments by the Federal Government and industry, educational and public information initiatives, and Federal and State tax incentives to meet the goal established under paragraph (1);
(5) consider the role that the use of hydrogen in stationary electric power applications, as well as advanced vehicle technologies, will play in stimulating the transition to hydrogen-fueled hybrid electric vehicles. Also consider whether other technologies would be less expensive or could be more quickly implemented than fuel cell technologies to achieve significant reductions in carbon dioxide emissions and oil imports;
(6) take into account any reports relating to fuel cell technologies and hydrogen-fueled vehicles, including (a) the National Academies report issued in 2004 entitled Hydrogen Economy: Opportunities, Costs, Barriers, and R&D Needs; and (b) the report prepared by the U.S. Fuel Cell Council in 2003 entitled Fuel Cells and Hydrogen: The Path Forward;
(7) consider the challenges, difficulties, and potential barriers to meeting the goal established under paragraph (1); and
(8) with respect to the budget roadmap (a) specify the amount of funding required on an annual basis from the Federal Government and industry to carry out the budget roadmap; and (b) specify the advantages and disadvantages to moving toward the transition to hydrogen in vehicles in accordance with the timeline established by the budget roadmap.
(9) Write a report documenting its study and assessment.

The committee's priorities from the statement of task were as follows:

- Establish as a goal the **maximum practicable percentage** of vehicles that can be fueled by hydrogen by 2020;
- Determine the funding, public and private, to reach that goal;
- Determine the government actions required to achieve the goal;
- Establish a budget roadmap to achieve the goal;
- Evaluate the synergy between the use of hydrogen in stationary electric power applications and in hydrogen-fueled vehicles; and
- Assess other technologies which could achieve significant CO_2 and oil reductions by 2020.

This report builds on an earlier NRC report prepared for DOE, *The Hydrogen Economy: Opportunities, Costs, Barriers, and R&D Needs* (NRC, 2004). That study examined how the hydrogen economy might work and the track the nation might follow, especially with regard to research and development (R&D), to get there. This report updates the review of the rapidly improving technology and focuses on the resources needed to have a significant number of HFCVs in operation by 2020. Increasing concern over both petroleum imports and climate change suggests that an earlier introduction of HFCVs, relative to the projections in the 2004 report, could be desirable. In assessing the potential of HFCVs to achieve significant reductions in oil imports and CO_2 emissions, the committee also benefited from interactions with the NRC's Committee on Review of the Research Program of the FreedomCAR and Fuel Partnership, which evaluated progress in the R&D programs related to fuel cell vehicle technologies and hydrogen production technologies in the partnership between DOE, the three domestic automotive companies, and five energy companies (NRC, 2008). Many of the technical conclusions in this report stemmed from that cooperation. The present committee also conducted its own analysis on a variety of issues and heard from many experts in industry, academia, and research centers, as listed in Appendix B.

In this report:

- Chapter 2 presents a framework for considering an accelerated transition to hydrogen fuel cell vehicles.
- Chapter 3 then reviews the progress of the technologies that will be necessary to make the introduction of HFCVs feasible. This includes hydrogen production as well as the vehicles themselves.
- Chapter 4 examines two alternative approaches to reducing the use of petroleum and emissions of carbon dioxide from vehicles—advanced conventional vehicles and fuels derived from biomass (biofuels)—to compare their potential to HFCVs.
- The possible relationship between hydrogen for transportation and the electric power system is explored in Chapter 5. This includes an examination of electrolysis for hydrogen production during the transition, as well as central station plants producing hydrogen that could also be used for stationary electricity production.
- Chapter 6 presents several scenarios designed to analyze the transition and its impacts. Building on the technology assessment in Chapter 3, one scenario examines the rapid introduction of HFCVs. Two others look at the rapid improvement in conventional fuel economy and at the use of biofuels, as discussed in Chapter 4. Finally, another scenario projects the impact on fuel use and CO2 emissions if all paths are pursued simultaneously. Variants of these scenarios also were analyzed, as reported in Appendix C.
- The funding for research, development and demonstration (RD&D), as well as the investments in vehicle production and hydrogen infrastructure that will be required for a transition to HFCVs, are discussed in Chapter 7. The chapter culminates with the budget roadmap mandated in the statement of task above, plus a discussion of how costs might be shared between the private sector and government.

• The federal government must supply more than just funding to stimulate a transition to fuel cell vehicles. A variety of other policy requirements and options are explored in Chapter 8.

• Finally, Chapter 9 assesses the broad advantages and disadvantages of accelerating this transition.

REFERENCES

NRC (National Research Council). 2004. *The Hydrogen Economy: Opportunities, Costs, Barriers, and R&D Needs*. Washington, D.C.: The National Academies Press.

NRC. 2008. *Review of the Research Program of the FreedomCAR and Fuel Partnership: Second Report*. Washington, D.C.: The National Academies Press.

2

Toward a Substantial and Durable Commitment: The Context of the Study

This chapter sets out the worldview and philosophy that guided the committee in responding to the inquiries of the 2005 Energy Policy Act. The multiple questions posed by the statement of task should be understood within the context of the committee's overall mission—to assess the resources the United States would need to support a transition in motor vehicles, fuels, and fueling infrastructure aimed at accomplishing three essential public goals:

1. Reduce the nearly complete dependence of road transportation on petroleum in order to improve energy security in the face of political instability among oil producers and mitigate the eventual peak in conventional oil production;
2. Lower the emissions of greenhouse gases from motor fuel production and use in order to sharply reduce the impact of motor vehicle use on the global climate; and
3. Maintain economic competitiveness and growth while achieving the first two goals.[1]

The issues underlying these goals are large, persistent, and global. They will not yield to a quick fix, nor can they be addressed independently. Successful policy must deal with them as a whole, which requires supporting a balanced portfolio of technology options rather than emphasizing a single solution.

Building such a portfolio can diversify the risk of delay or even failure of any one technology. More important, a portfolio can deliver benefits throughout the lengthy period (perhaps extending to 2050) required for a hydrogen-based transportation system to mature. Consider as examples the improved fuel economy of conventional vehicles, the intent of the newly revised CAFE (corporate average fuel economy) standards; hybrid electric vehicles also offering significantly improved fuel economy; and motor fuels derived from biomass.

These options can considerably reduce oil consumption over the next 20 years, but they are unlikely to eliminate the problems of oil dependence and climate change. Their ultimate resolution will require bringing to market vehicle technologies such as the hydrogen fuel cell vehicle (HFCV) or fully electric vehicles.[2] Yet these are unlikely to enter the market in sufficient numbers over the next 20 years to substantially reduce petroleum consumption. Thus, a technology portfolio that includes all of these options will deliver greater benefit across the intervening 20 or so years. Nevertheless, hydrogen technologies and infrastructure offer the potential, once successfully developed, to achieve fully the threefold goals of energy policy—hence, their emphasis in the congressional inquiry and in the committee's response to it.

Initiating a fundamental energy transition will require a policy commitment on the part of the federal government. This commitment and the policies that implement it must remain substantial and durable over the decades needed to complete the transition:

- *Substantial*, in that policy provides meaningful incentives for fuel economy where the market price of the fuel does not include externalities, such as environmental and health costs from emissions or an oil vulnerability premium; and

[1]The committee was not asked to study economic competitiveness and growth but notes them as vital considerations in the policy process: to the extent that remedies to energy security and climate change impose near-term economic burdens, their implementation will be more difficult. Energy security and climate change are discussed in greater detail in this chapter. Economic issues are of concern throughout the report.

[2]This judgment assumes that hydrogen and electric energy can be made in a way that does not release greenhouse gases over the long term. Some transition strategies, for example, could allow the manufacture of hydrogen from natural gas or electricity from coal without capture and sequestration of the carbon emitted in those processes. The greater efficiency of the fuel cell or electric vehicle, relative to the conventional vehicles that they displace, would offset the carbon release during a transition. However, in a mature hydrogen economy, effective capture and sequestration of the carbon dioxide would become essential.

- *Durable*, in that policies remain in place long enough for consumers, entrepreneurs, technologists, and investors to make the needed commitments of their own time and resources.

To the extent that the United States makes such a commitment, the history of other technology transitions shows that our market-based economy and others around the world will prove highly effective in achieving the public goals of energy security and climate stabilization while preserving healthy and sustainable economic growth.

ENERGY SECURITY

The issue with energy security arises chiefly from the near-total dependence[3] on conventional petroleum as the source of fuel for the transportation sector in the United States and most of the world's economies. Adverse consequences arise from global dependence on petroleum from regions of the world that are either unstable or inimical to U.S. interests.[4] Insecurity in petroleum supply holds the prospect for large-scale disruptions of the world economy. Energy insecurity is likely to increase over time as a result of the following:

- The prospect of disruption of the petroleum supply chain, through terrorist attack, political instability in the supplying nations, or natural disaster;
- Projected demand growth, especially among the developing nations of non-OECD (Organization for Economic Co-operation and Development) Asia (about 2.7 percent per year until 2030), which strains reserve production capacity that might have offset such disruptions (EIA, 2007); and
- The possibility that conventional oil production may peak much sooner than accounted for in business-as-usual forecasts.

The current petroleum market lacks the excess production capacity that characterized past decades, and production and demand remain in close daily balance. This means that any disruptive event, whether from a natural disaster or terrorist activity, can cause severe and lasting price shocks, leading to worldwide economic dislocation.

This situation is unlikely to improve in the near future. Demand continues to increase at the same time that conventional petroleum production faces a leveling and/or peaking of world oil production. In a recent study, the U.S. Government Accountability Office noted that "the total amount of oil underground is finite, and, therefore, production will one day reach a peak and then begin to decline. Such a peak may be involuntary if supply is unable to keep up with growing demand" (GAO, 2007, p. 6). Similarly, the International Energy Agency (IEA) concluded, "Worldwide, the rate of [oil] reserve additions from discoveries has fallen sharply since the 1960s. In the last decade, discoveries have replaced only half the oil produced" (IEA, 2006, p. 132).

The literature offers a wide range of estimates concerning the timing of a maximum in world oil production because the data needed for more precise forecasting are not widely available. Much useful information is (1) proprietary to companies, (2) a state secret in the major oil exporting countries, and/or (3) biased to achieve political and economic objectives.

For example, a recent study by the National Petroleum Council stated that "there are accumulating risks to continuing expansion of oil and natural gas production from the conventional sources relied upon historically. These risks create significant challenges to meeting projected energy demand." These risks are both geological and geopolitical. Further, "Forecast worldwide liquids production in 2030 ranges from less than 80 million to 120 million barrels per day, compared with current daily production of approximately 84 million barrels. The capacity of the oil resource base to sustain growing production rates is uncertain" (NPC, 2007, p. 91).

To be sure, enormous resources of unconventional oil—for example, oil shale or coal in the United States and tar sands in Canada—could be liquefied and substituted for oil. Exploiting these resources could greatly extend the availability of gasoline and diesel fuel, but would also raise environmental issues. Chiefly, they would nearly double the carbon dioxide (CO_2) emitted per gallon of fuel consumed, unless the emissions from production can be captured and permanently sequestered, and their use would increase the demand for water.

In addition, a peaking or leveling in production would probably be attended by price increases, and these would induce a demand response—some combination of (1) greater efficiency in converting petroleum to services and (2) simply doing without. However, examining the potential contribution of either unconventional fuel resources or demand response falls outside the committee's assigned tasks, and they are not considered further here.

CLIMATE CHANGE

The second element of the energy "trilemma" concerns the environmental consequences of the buildup of CO_2 and other greenhouse gases in the atmosphere.[5] Light-duty vehicles generate one-third of global CO_2 emissions and about a third of U.S. emissions. Capturing CO_2 emissions from individual vehicles is effectively impossible, so reduc-

[3]In the United States, 96 percent of the primary energy used in transportation comes from conventional petroleum (EIA, 2007, Table 2.1e, p. 42).

[4]See, for example, Council on Foreign Relations, 2006, *The National Security Consequences of U.S. Oil Dependency*, Independent Task Force Report No. 58, Washington, D.C.

[5]In addition to carbon dioxide, the "greenhouse gases" generally include water vapor, hydrogen itself, nitrous oxide, methane, hydrofluorocarbons (HFCs), perfluorocarbons (PFCs), and sulfur hexafluoride.

tions in the transportation sector can be effected only by improved fuel economy and/or replacement of current fuels with lower-carbon or zero-carbon fuels. Hydrogen contains no carbon at all, but the production processes currently available emit CO_2—either from natural gas and other fossil fuels used to manufacture hydrogen or from fossil fuels that generate the electricity used to make hydrogen via electrolysis. Even including these production-derived carbon emissions, however, hydrogen fuel cell vehicles can reduce the well-to-wheels carbon given off by light-duty vehicles because of the greater efficiency of the fuel cell.

Nevertheless, achieving deep reductions in emissions from hydrogen production will require development and use of processes that can capture and sequester the CO_2 generated in hydrogen manufacture, as well as greater use of low-carbon or zero-carbon energy sources for electricity generation. Biofuels, especially if produced renewably, also would reduce carbon emissions relative to conventional fuels.

Although long-term in consequence, the threat of global warming is of immediate concern, because moderate actions taken now could preclude the need for drastic actions taken later. According to the world's clearinghouse for peer-reviewed climate science, the Intergovernmental Panel on Climate Change (IPCC), "The global atmospheric concentration of carbon dioxide has increased from a pre-industrial value of about 280 parts per million (ppm) to 379 ppm in 2005. The atmospheric concentration of carbon dioxide [and methane] in 2005 exceeds by far the natural range over the last 650,000 years (180 to 300 ppm) as determined from ice cores" (IPCC, 2007, p. 2).

In 2005, a National Research Council (NRC) report focused on these conclusions, stating that "in the judgment of most climate scientists, Earth's warming in recent decades has been caused primarily by human activities that have increased the amount of greenhouse gases in the atmosphere" (NRC, 2005, p. 2). Although the debate over the science of historical climatic changes has been largely resolved and there is agreement about the potential influence of continued greenhouse gas emissions on climate, the 2005 NRC report notes that "there is still legitimate debate regarding how large, how fast, and where these effects will be" (p. 2).

Most recently, in 2007, the IPCC wrote that "[w]arming of the climate system is unequivocal, as is now evident from observations of increases in global average air and ocean temperatures, widespread melting of snow and ice, and rising global average sea level" (IPCC, 2007, p. 5). The warming has been especially acute since 1995. According to the IPCC, "eleven of the last twelve years (1995-2006) rank among the 12 warmest years in the instrumental record of global surface temperature (since 1850)" (p. 5).

With regard to the consequences of the greenhouse gas buildup, the 2007 IPCC report noted that climate change risks are likely (greater than 66 percent) to include droughts, sea level rise, and increased tropical cyclone activity. Increased heat waves and heavy precipitation events, which can lead to flash floods and severe erosion, are very likely (greater than 90 percent).

The committee has not assessed climate change risks but concludes that if immediate action is required to reduce CO_2 emissions, the transportation sector could provide a significant share of the reductions. The hydrogen technologies discussed in this report are particularly promising for large-scale reductions over the longer term.

MOTIVATING THE PRIVATE SECTOR TO MAKE THE ENERGY TRANSITION

The problems that arise from the security-environment-economy trilemma become manifest to the public chiefly as motor fuel prices, trade imbalances, defense expenditures, and inflationary pressures and less visibly but more consequentially, threats such as worldwide economic instability, foreign policy challenges, and eventually global climate change. For the past 35 years, proposed solutions have tended to emphasize one or, at most, two of these, with neglect of the others. Energy policy has suffered from such selective inattention because the way in which one part of the problem is addressed strongly influences the other parts. Genuine progress requires a portfolio solution and a substantial commitment that remains durable over the 40 or so years needed for a transition. Such a solution is unlikely to arise from any linear summation of the solutions for each component.

In addition, near-term solutions should be considered in the context of long-term policy goals. For example, a more rapid transition might be achieved with use of technologies that would not fit well in a mature, post-transition energy economy—for example, the venting of CO_2 from distributed production of hydrogen from natural gas and perhaps even from some large-scale production might be tolerated to speed a transition, provided that some means of eventual carbon capture and sequestration could reasonably be ensured. Nevertheless, the sum of such short-term solutions is unlikely to lead to the most efficient or desirable long-term solution.

The committee believes that energy policy can no longer afford the luxury of short-term thinking or of selective inattention.

In a free society, the government cannot command an energy transition into being, and so must engage the private sector. Indeed, government investment in the energy transition will ultimately prove small in comparison with that from private sources—individuals, entrepreneurs, investors, and businesses. For these reasons, the committee was asked to include private resource commitments in its estimates.

To effectively marshal private resources, energy policy must create the appropriate framework of regulation and incentive, augmented with meaningful investments in the research, development, and demonstration (RD&D) with greatest leverage for the transition. With a wise and effective framework in place, private resources will flow to large and durable opportunities thus created.

The history of large-scale transformations in other industries can illuminate the relative advantage of each in accelerating the energy transition. In general, two distinct kinds of economic activity operate in parallel to set the pace and direction of change. The first is a process of *evolutionary* change in which improvements in technologies and infrastructure now in place or emerging in the marketplace dominate progress. The cumulative effect of these incremental improvements can be striking—examples include the development of faster microprocessors by Intel or higher-resolution medical scanning devices by General Electric. Within the energy-fuels sector, the auto industry has improved the fuel efficiency of motor vehicles between 1 and 2 percent each year for several decades. However, since the early 1980s, automakers have also responded to consumer demand by using these efficiency gains to increase vehicle weight and performance, rather than improve fuel economy.[6] Similar improvements can be shown in the efficiency of electric generation and conventional oil production. In general—though not exclusively—market incumbents enjoy an advantage in evolutionary technological change. Such evolutionary improvements provide immediate progress toward the policy goals, but also compete with the revolutionary technologies.

In contrast, *revolutionary* technologies, when successful, redefine the marketplace and the competitive environment. Progress comes through discontinuous change and offers far-reaching solutions as distinct from incremental improvements. Historic examples include mechanical refrigeration, solid-state electronics, and the telephone. The HFCV and the electric vehicle have the possibility to radically transform the worldwide motor vehicle industry, its attendant fuel infrastructure, and the electric utility industry. In general—and with many important exceptions—new ventures enjoy a competitive advantage in initiating radical technology change because they have no incumbent technologies to defend.

Policy can accelerate both radical and incremental change, but distinct instruments are needed for each. This implies that a portfolio approach will accelerate the energy transition most effectively. At any point in time, a well-founded energy policy would support a portfolio of improving, emerging, and potentially revolutionary technologies, and it would influence both established companies and entrepreneurial ventures.

PRINCIPLES FOR EFFECTIVE TRANSITION POLICY

Several general desirable characteristics will influence the success of any policies aimed at achieving the maximum practicable market share of vehicles fueled by hydrogen, the desideratum the committee was asked to estimate. The committee has used these principles throughout its report:

- Energy policy must offer greater *certainty and predictability* than at present, if private markets are to marshal the resources to accelerate the transition. Entrepreneurs, innovators, and larger industries can manage uncertainties in technology and markets. They do not, however, respond as effectively to political uncertainty.
- Policies should send *consistent* messages to innovators and investors, not only within the United States, but insofar as possible internationally.
- Policies must be *substantial*. Half-measures produce half-results.
- Policies must have *integrity*. Earmarked funding, for example, dilutes the resources available for essential research and demonstration; similarly, special exemptions for favored industries or protected groups erode the public sense that the pain and the gain are fairly shared.

The committee has condensed these principles into two core concepts—policies should be *substantial* and they should be *durable*. "Substantial" means that incentives are large enough to make a difference in marketplace decisions, and "durable" means that policy incentives remain in place long enough for innovators to respond—which might require a planned phaseout to ensure sustainability.

Contrasting cases of incentives for wind energy production and for solar photovoltaic technologies are used to illustrate these principles. These cases compare and contrast policy initiatives in two countries: (1) the consistent and long-term feed-in tariff program for solar energy in Germany; and (2) the long-term, but intermittently authorized, production tax credit for installation of wind energy capacity in the United States. Both of these programs were intended to promote the installation of renewable energy capacity, and both sought to build a viable long-term industry in their respective countries.

The committee has not examined the merits of the goals that either policy sought to achieve and makes no recommendations regarding their adoption. Rather, it uses these cases to illustrate the principles of substantial and durable energy policy and how these can become important in achieving the public purpose.

The Feed-in Tariff Experience in Germany

On April 1, 2000, the German government introduced the Renewable Energy Resources Act—known in Europe as the EEG—to provide substantial incentives for the installation of renewable resources connected to the electricity grid. After consideration of a wide variety of potential incentives, including a renewable portfolio standard (RPS), the German parliament devised an incentive program referred to around the world today as the feed-in tariff.

The EEG provided incentives to install alternative renewable options from biomass to wind energy, but by far the most aggressive incentive was granted to installation of photovol-

[6]See Figure 4.1 (Chapter 4).

taic (PV) systems. The PV feed-in tariff satisfies the criteria suggested in this report as appropriate for a policy initiative aimed at creating rapid and significant change in an existing energy delivery system:

- *Substantial.* The program offered incentives to installers of PV systems ranging from 45.7 to 57.4 euro-cents/kWh, many times above the wholesale and retail electricity rates throughout Germany.
- *Durable.* The contracts with PV system owners are for a 20-year period at a fixed payment rate. This allows the purchasers of a PV system to determine, subject only to variations in available sunlight, the cash flow from the investment made.
- *Sustainable.* No subsidy should endure indefinitely, and so a predictable phaseout becomes an important element of durability. This "digressive" tariff declines over time in a predictable manner. The tariff paid to a system owner for an installation brought on line at any time after January 1, 2002, is reduced by 5 percent from the rate paid for an installation made in the previous year, although the contract for that reduced rate was still for 20 years. The expectation underlying this digressive approach was that an expanding PV manufacturing base would lead to declining prices for delivered PV systems over time.

Utility customers pay for the incentive through the established rate structure, and the utility in turn makes payment directly to the owner of the PV system connected to its grid. Thus, the feed-in tariff becomes a form of tax, collected by the utility, much in the same way that certain weatherization programs and special low-income rates are financed by U.S. utilities via a small charge on each kilowatt-hour of electricity delivered.

The architects of the EEG were confident that the feed-in-tariff policy would "create a stable investment climate, while RPS policies would not" (Rickerson and Grace, 2007). Indeed the UK Treasury's recently published *Stern Review* noted that "feed-in mechanisms achieve larger deployment [of incented resources] at lower costs" because of the "assurance of long-term price guarantees" (Stern Review, 2006, p. 366). The *Stern Review* goes on to say that other types of incentives are less successful and more costly because "uncertainty discourages investment and increases the cost of capital as the risks associated with uncertain rewards require greater [return] rewards" (p. 366).

The EEG feed-in tariff has created a large, global, solar energy market in Germany, spawning numerous rapidly growing companies that did not exist before the EEG was passed. These new enterprises grew in Germany (QCells, for example) and around the world (SunTech in China, for example). Today, 18 countries in the European Union have adopted feed-in tariffs to promote the deployment of renewable generation technologies. For example, the feed-in tariffs initiated in Greece in June 2006 are 40-50 euro-cents/kWh, depending on system size and mainland versus island location.[7]

The committee has not studied whether a feed-in-tariff approach would mesh well with the requirements for rapid deployment of hydrogen systems and makes no recommendations regarding this policy instrument. Yet whatever policies are ultimately adopted in the United States, consideration of the basic principles employed by the German program will surely lead to more rapid and effective results in establishing hydrogen policy that proves:

- Substantial enough to influence marketplace decisions;
- Durable for long enough to stimulate innovation, investment, and cost reduction by suppliers of hydrogen vehicles and infrastructure technology, and
- Sustainable by providing for reduction of the incentive over time, as a signal to the market that hydrogen systems must become fully competitive with alternatives over a well-defined period.

Wind Energy Production Tax Credits

The solar feed-in tariff program in Germany stands in marked contrast to the production tax credit (PTC) program for wind energy in the United States. The intent of the PTC was to reduce the cost of wind energy and thereby make it more attractive to electric utilities and investors. PTCs for wind energy were first established in the 1992 Energy Policy Act and were initially valued at 1.8 cents/kWh produced during the first 10 years of operation. This credit applied to installations put in service from 1994 through June 30, 1999. The PTC has been extended periodically, in similar form, and is still available today.

However, the intermittency of this policy has inhibited the full achievement of its goals. Since its initial 5½ years, the PTC has never been renewed for more than 2 years, and it was allowed to lapse completely on three separate occasions for periods ranging up to several months, as shown in Table 2.1. Also, although in some cases authorization was retroactive after a lapse, this provided little incentive because many wind projects needed the financeable future income stream from the PTC to be economically viable.

[7]It is possible that the feed-in tariff has been too successful. PV costs have not declined as much as expected, largely because soaring demand for PV-grade silicon has increased its cost. The German government is considering increasing the rate at which the guaranteed price drops for new installations, but such changes remain speculative as of this writing. The committee notes that these proposed changes would reduce the subsidies offered to new participants, not because the feed-in tariff has proved ineffective, but rather because policy officials seek to direct the subsidies to other forms of renewable energy, especially wind. The core idea that substantial and durable incentives can make a difference remains much in evidence—witness the unlikely emergence of Germany as an economic power in solar photovoltaics.

TABLE 2.1 Legislation of Production Tax Credits for Wind Energy in the United States

Date of Legislation	In-service Dates
1992 Energy Policy Act	1994-June 30, 1999
PTC lapsed	June 30, 1999-December 1999
December 1999	2000-2001
PTC lapsed	January 2002-February 2002
February 2002	2002-2003
PTC lapsed	January 2004-October 2004
October 2004	2004 (retroactive)-2005
August 2005	2006-2007
December 2006	2008

SOURCE: Courtesy of American Wind Energy Association.

The intermittent nature of the PTC has retarded both the installation of wind capacity in the United States and the development of the wind energy industry in the United States. Figure 2.1 shows annual additions of wind power in the United States, with telling decreases in installations in 2000, 2002, and 2004, following the expiration of the PTC in 1999, 2001, and 2003. More significantly, in recent years (2005 and 2006) the PTC has been extended before it expired, and steady growth in wind capacity additions is anticipated during the now authorized 2005-2008 period.

Although the impact of the PTC lapses on installation of wind capacity can easily be seen in Figure 2.1, it is harder to evaluate directly the damage done to U.S. wind energy businesses. Many of the impacts of the boom-bust cycle are difficult to assess and even more difficult to value. Inefficiencies in production, costs of holding inventory, difficulties in managing supply chains, and costs associated with maintaining (or reducing) a workforce all represent very real, but hard-to-see, costs to a business trying to navigate the uncertain market illustrated in Figure 2.1. Furthermore, many companies may simply never have evaluated their losses, or even if these numbers were estimated, they are not generally available publicly.

ENTREPRENEURSHIP AS A FORCE FOR CHANGE

A free economy grows and adapts through competition, the "creative destruction" used by economist Joseph Schumpeter (1883-1950) to describe the industrial and societal transformations that accompany widespread innovation. The central participants in this innovation process include (1) established for-profit companies offering products and services pertinent to the energy transition; (2) new entrepreneurial ventures just emerging into the marketplace; and (3) investment institutions providing capital to both. Each has a unique role in the process of creative destruction, and transition policies must reach all three to motivate change and fully realize the opportunities for economic growth.

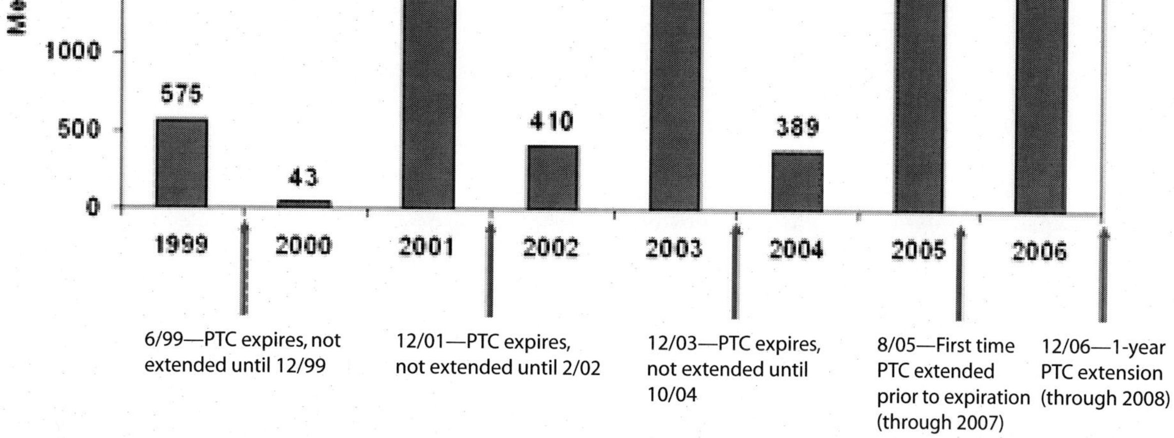

FIGURE 2.1 U.S. wind power capacity additions, 1999-2006. SOURCE: Courtesy of American Wind Energy Association.

Nevertheless, energy policy has underappreciated the role of the entrepreneurial venture. This committee seeks to recall attention to it because the pace of the energy transition and the resources needed to achieve it will be influenced strongly by the success of these entrepreneurial ventures.

Entrepreneurial Companies in the Transition to Hydrogen

In many sectors of the economy, entrepreneurial venture capital-backed companies have been prominent players in driving toward a new paradigm. Certainly this has been true since deregulation of the telecom industry, which led to the emergence and rapid growth of many new, and now very large, companies such as Cisco, Palm, Ciena, Nokia, and the like. The same has occurred in the computer and software sector, with entrepreneurial companies such as Apple, Compaq, Microsoft, Sun Microsystems, Dell, and many others now dominant in their fields.

In the economy as a whole, entrepreneurial companies with venture capital backing have had an enormous impact. According to a study issued recently by the National Venture Capital Association, venture-backed companies employed 10.4 million people and generated $2.3 trillion in revenue in 2006, which represented 9.1 percent of the total private sector work force and 17.6 percent of the total GDP (National Venture Capital Association, 2007, p. 5). These same companies outperformed the overall economy by 2:1 in both rate of job growth and sales growth.

Yet in the field of hydrogen production infrastructure and applications it has been widely assumed that the transition, when it occurs, will be led by the major automotive and energy companies, and little attention is given to the innovation being achieved by small entrepreneurial companies. Indeed, most of the major automotive players worldwide do have active fuel cell vehicle programs, and oil companies, notably Shell, Chevron, and BP, as well as several of the large industrial gas companies, have substantial activities in hydrogen production and infrastructure development. Large industrial and aerospace companies, including United Technologies and General Electric, also have efforts under way in stationary fuel cell systems and hydrogen production technologies.

It is difficult to determine how much capital has been invested by major corporations around the world in hydrogen technology because programs in these companies are typically conducted by divisions or operations that do not issue financial reports separately from the parent. Estimates of expenditures based on anecdotal evidence provided informally by corporate executives range from $2 billion or $3 billion to well over $10 billion.

Despite the evidently substantial activities of large corporate players in the hydrogen field, it is not by any means certain that the leaders in the transition will be the large companies that are currently active. Indeed many entrepreneurial companies are pursuing hydrogen-relevant technologies of all types—fuel cells, transition-scale hydrogen production, and on-vehicle hydrogen storage. Any number of those might emerge as leaders in the hydrogen transition, much as happened in the telecommunications and computer fields.

Over the past decade, many of these entrepreneurial venture-capital-backed companies were able to complete public offerings on the NASDAQ market, mainly in 1999-2000. As shown in Table 2.2, the total capital raised by the 10 NASDAQ-listed public companies was nearly $4 billion, as of year end 2006. It is interesting to note that the financial statements for these companies show that $374 million of additional capital was raised by this group of companies in 2006, more than half of it resulting from an investment in Plug Power by a Russian consortium.

In addition, six small companies involved with hydrogen systems are listed on the London AIM exchange with total invested capital of more than $322 million (see Table 2.3).

TABLE 2.2 Capital Invested in Selected Small Public Hydrogen and Fuel Cell Companies Listed on the NASDAQ

Company[a]	Invested Capital (million dollars) (12/31/05)[b]	(12/31/06)[b]
Ballard Power Systems (BLDP)	1,161	1,170
Distributed Energy Systems (DESC)	221	236
Fuel Cell Energy (FCEL)	530	531
HOKU Scientific	32	33
Hydrogenics (HYGS)—includes Stuart Energy	319	321
Mechanical Technology (MKTY)	122	131
Medis Technologies (MDTL)	209	287
Millennium Cell (MCEL)	108	114
Quantum Fuel Systems (QTWW)	255	288
Plug Power (PLUG)	532	752
Total	3,489	3,863

[a]NASDAQ only, excludes AIM listed companies.
[b]Or nearest year end.
SOURCE: Compiled by the committee from publicly available financial statements on www.edgar-online.com (accessed November 2007).

TABLE 2.3 Capital Invested in Selected Small Hydrogen and Fuel Cell Companies Listed on the AIM Market in the United Kingdom

Company	Invested Capital Local Currency	US$
Ceres Power (CWR)	£18,062,000	35,943,380
Ceramic Fuel Cells, Ltd (CFU)	AU$185,549,893	154,006,411
ITM Power (ITM)	£12,702,049	25,277,078
Protonex (PTX)	$29,609,155	29,609,155
PolyFuel (PYF)	$58,673,829	58,673,829
Voller (VRL)	£9,343,000	18,592,570
Total		322,102,423

SOURCE: Compiled by the committee from publicly available financial statements on the Internet, accessed July 2007.

TABLE 2.4 Capital Raised by Private Sector Entrepreneurial Companies for Hydrogen Technologies

Area of Company Focus	Number of Companies in Database	Capital Raised 2002-2007 ($)
Fuel cells (of all types)	38	468,107,388
United States	21	327,595,800
Europe	13	91,212,248
Canada	4	49,299,340
Fuel cell components (e.g., membranes)	8	127,490,300
United States	2	59,900,000
Europe	6	67,590,300
Hydrogen production (e.g., electrolyzers, reformers)	12	64,688,900
United States	6	47,255,000
Europe	2	3,800,000
Canada	4	13,633,900
Hydrogen storage (e.g., tank, hydrides)	3	6,710,309
United States	2	808,000
Europe	1	5,902,309
Hydrogen infrastructure (e.g., compressors, dispensers)	1	2,055,000
Canada	1	2,055,000
Other (e.g., integrators, applications providers, vehicle retrofit)	6	135,786,000
United States	3	93,020,000
Europe	2	11,000,000
Canada	1	31,766,000
Total	68	804,837,897

SOURCE: CleanTech Network, personal communication, July 16, 2007.

Two of these companies are U.S.-based, one is Australian, and three are in the United Kingdom.

It is much more difficult to obtain data on investments in privately owned companies that pursue hydrogen technology because they do not make balance sheet data available publicly. Data received from the U.S. Fuel Cell Council and from two venture funds that were willing to share their deal logs yield some approximate estimates: between 104 and 160 private entrepreneurial companies in the United States, Canada, and Western Europe are engaged in activities related to hydrogen and fuel cells. Almost no reliable data are available about private companies in the rest of the world that are pursuing hydrogen developments. Probably the best source of data on such companies is the Cleantech Network; it has graciously provided aggregated data that show capital invested in more than 68 private fuel cell and/or hydrogen companies to be almost $805 million since 2002 (Table 2.4), an astonishingly high number in view of the very difficult investment climate in the early part of the decade.

The 68 companies that reported financial investment data to the Cleantech Network were fewer than half of the 147 companies that the Network has identified in the hydrogen fuel cell space, so it is possible that the actual amount invested is more than double the total shown in Table 2.4.

Among the private and small public companies engaged in work on hydrogen may be a few that will be the future Ciscos and Microsofts of the field, when and if the transition to hydrogen occurs. The challenge for investors is, of course, to anticipate which ones will achieve long-term success. Those who have studied the emergence of new technologies, such as radio and automobiles, point out that initially there were hundreds of new entrants but, in the end, only a few of those companies survived and thrived.

As in all fields, entrepreneurs interested in hydrogen as a business opportunity will respond vigorously to clear signals about market opportunity. Deregulation of the telecommunications industry, for example, created a flood of new entrants, pursuing opportunities that had previously been denied to all but AT&T. Virtually all hydrogen-related companies saw increases in their share prices when the FreedomCAR program was announced, demonstrating a vigorous response to government signals. During 1998-2001, signals from the auto industry about imminent introduction of hydrogen-fueled vehicles led to a surge in private and public capital flowing into entrepreneurial companies that offered technologies able to serve this anticipated new market.

Since early 2000, however, the market has come to realize that the technology development timetables in hydrogen are longer than many had thought and the costs to achieve acceptable price and performance are greater than originally anticipated. As a result, in 2007 there was little investor enthusiasm for investment in hydrogen (except for the Russian consortium said to be investing in Plug Power). As a result, there are few new public offerings (and some that came to market had to be withdrawn) and little interest in supporting new private companies in the space. Even companies with successful track records in development and market entry are facing challenges in raising the capital needed to keep moving forward.

Interest could be reignited quickly if the United States were to make a substantial and durable commitment to policies that would send a clear signal to the capital markets that hydrogen was a priority option in addressing the security-environment-economy trilemma described earlier in this chapter.

CONCLUSION

The introduction of production hydrogen fuel cell vehicles and the establishment of a hydrogen infrastructure are high-risk, high-payoff endeavors that would promote the global good—through reduced oil dependence and risk of climate change. Major automotive companies, fuel companies, and entrepreneurial ventures have devoted considerable resources to developing such vehicles, with support from governments around the world. Nevertheless, difficult technical issues remain, especially the development of cost-effective hydrogen storage on the vehicles and fuel cell systems that meet durability and customer requirements. For this reason, all of the private companies that are involved will continue to depend on a government commitment that is substantial enough to make a difference and that remains durable until HFCVs are competitive.

REFERENCES

EIA (Energy Information Administration). 2007. International Energy Outlook 2007. IEO 2007 Reference Case, p. 30. Washington, D.C.

GAO (Government Accountability Office). 2007. Crude Oil—Uncertainty About Future Oil Supply Makes It Important to Develop a Strategy for Addressing a Peak and Decline in Oil Production. GAO-07-283. Washington, D.C.

IEA (International Energy Agency). 2006. World Energy Outlook 2005. Washington, D.C.

IPCC (Intergovernmental Panel on Climate Change). 2007. Summary for Policymakers. In Climate Change 2007: The Physical Science Basis. Contribution of Working Group I to the Fourth Assessment Report of the Intergovernmental Panel on Climate Change. Cambridge, UK: Cambridge University Press.

NPC (National Petroleum Council). 2007. *Facing the Hard Truths About Energy—A Comprehensive View to 2030 of Global Oil and Natural Gas*. Washington, D.C.

NRC (National Research Council). 2005. *Review of the Research Program of the FreedomCAR and Fuel Research Partnership: First Report*. Washington, D.C.: The National Academies Press.

NRC. 2006. *Understanding and Responding to Climate Change: Highlights of National Academies Reports*. Updated in 2006. Washington, D.C.: The National Academies Press.

National Venture Capital Association. 2007. *Venture Impact: The Economic Importance of Venture Capital Backed Companies to the U.S. Economy*. Fourth Ed. Arlington, Va.

Rickerson, W., and R.C. Grace. 2007. The Debate over Fixed Price Incentives for Renewable Electricity in Europe and the United States: Fallout and Future Directions. White Paper prepared for the Heinrich Boll Foundation.

Stern Review. 2006. *Policy Responses for Mitigation: Accelerating Technological Innovation*. Cambridge, UK: Cambridge University Press.

3

Hydrogen Technology

This chapter reviews the hydrogen production and fuel cell vehicle technologies that will be needed to establish a budget roadmap to achieve the "maximum practicable" goal spelled out in Chapters 6 ("U.S. Carbon Dioxide (CO_2) and Oil Reductions from Hydrogen and Alternatives") and 7 ("A Budget Roadmap," which covers the R&D, vehicle, and infrastructure costs entailed in moving to a self-sustaining hydrogen-based system). It details how the hydrogen production and delivery infrastructure might grow to meet the demand for hydrogen fuel and lays out a picture of how the future hydrogen infrastructure could evolve to support the growing number of hydrogen fuel cell vehicles (HFCVs), referred to as the maximum practicable rate (MPR) of penetration, developed and discussed in Chapter 6. In the MPR case, HFCVs are introduced starting with a few thousand in 2012, growing to a fleet of 2 million in 2020, 60 million in 2035, and 220 million in 2050.

In assessing these developments, the committee analyzed hydrogen production, delivery, and dispensing technologies and evaluated fuel cell and hydrogen storage. In doing so, its members met representatives of major hydrogen fuel and auto companies that have announced hydrogen programs, participated in studies of the National Research Council (NRC) Committee on the Review of the FreedomCAR and Fuel Partnership. (To allow for greater interaction, the two committees shared two members.) It also met with U.S. Department of Energy (DOE) officials to review its technical programs, conducted a detailed review of the literature with the help of a consultant, and reviewed publicly available sources that have a history of technology evaluations and/or product introductions.

Resource calculations in Chapters 6 and 7 include costs of both the HFCVs and the hydrogen infrastructure (for production, delivery, and dispensing), and this chapter includes estimated cost data for key technologies. The cost of the hydrogen fuel cell and onboard hydrogen storage system on the HFCV will be critical to its competitiveness with alternative vehicle technologies described in Chapter 4. The cost and technical status of hydrogen producing and delivering technologies from various primary feedstocks determined which technologies were included in the roadmap. This chapter focuses on the least expensive, most fully developed technologies with potential for low well-to-wheels (WTW) carbon dioxide (CO_2) release that are the most practical for different stages of the move to hydrogen. Using the criteria of cost and stage of technological advancement in this screening process effectively ruled out some technologies that are in early development phases or have high current costs but may have good potential for improvement. This fact does not imply that these technologies are unimportant or will not play roles in the future. The committee's task included estimating the costs of building the infrastructure to meet growing hydrogen demand during the transition to a self-sustaining hydrogen transportation system (i.e., one without subsidies). The technologies chosen represent just one scenario of what could happen.

This chapter addresses the readiness of the hydrogen production and delivery and HFCV technologies to begin to meet the MPR case. Much of the current development of HFCVs is aimed at the year 2015 for the most crucial early-stage technologies to be at the point at which decisions to move to mass commercialization could be made. This focus is consistent with the MPR case.

HYDROGEN PRODUCTION AND DELIVERY

Hydrogen can be produced from various resources, either in small facilities at the point of sale or in larger facilities farther away and requiring delivery. Getting to a self-sustaining market will require an evolution of the supply infrastructure to ensure the lowest possible costs and fuel availability at all times. The long-term vision of the hydrogen production and supply infrastructure is to produce large amounts of hydrogen from domestic resources with low WTW CO_2 emissions and deliver this to customers at a cost that is competitive with gasoline on a cost-per-mile traveled basis. The report

The Hydrogen Economy (NRC, 2004) indicated that this long-term potential could be possible once a large number of HFCVs are in the market and a mature hydrogen industry is in place. This section discusses the transitional period and lays out a picture of what this might look like. The technologies discussed are those that are the most well developed and nearest to technical and economic readiness. These are the technologies that are included in the resource estimates in Chapter 6, along three alternative time frames:

1. Early hydrogen delivery for first vehicle owners—supplied from existing industry excess capacity or small skid-mounted production appliances at existing gasoline stations
2. Midtransition vehicle owners—supplied mainly by on-site production at full-size refueling stations
3. Late transition to self-sustaining hydrogen transportation system—supplied mainly in large central production facilities and delivered by pipeline to refueling stations

Early Hydrogen Infrastructure

Before HFCVs can be marketed generally, there must be enough fueling facilities to convince buyers that fuel will be available when and where they need it. Based on the analogy with diesel fuel, between 10 and 25 percent of stations providing an alternate fuel gives customers confidence in fuel availability. Meeting this standard for early hydrogen vehicle users will be very challenging because hydrogen fueling equipment is much more expensive than traditional liquid fuels. Reducing the number of locations while still providing good coverage will be important to hold down capital costs and to increase the hydrogen volume sold at each location. Recent estimates (Nicholas et al., 2004) are that as few as 5 percent of stations offering hydrogen might satisfy customer concerns if their locations are closely coordinated to provide adequate coverage of urban and suburban areas. This very close coordination in locating stations is a new concept compared with today's free market and individual company decisions to determine locations. This concept involves coordination between auto companies, energy providers, and local governments that is not practical today because of antitrust concerns.

Convincing hydrogen suppliers to build hydrogen stations before the introduction of many HFCVs is sometimes referred to as the "chicken-and-egg problem." Hydrogen suppliers are reluctant to invest large sums before they know that many HFCVs will be sold. Similarly auto companies will not be able to sell many HFCVs without an adequate number of hydrogen fueling stations. A way around this quandary is to stage HFCV introduction in phases by region. This approach is referred to as the "lighthouse concept" (Gronich, 2007). For example, if the HFCV is first introduced in the Los Angeles area, then only 5 percent of the stations in that region need to offer hydrogen to provide adequate coverage. As more HFCVs are sold, the infrastructure expands, eventually to mid- and late transition supply options discussed in the next two sections. This is then repeated in more cities. This concept is more fully explained in Chapter 6, as is a list of possible cities.

The demand for hydrogen at these stations will be very low for the first several years of operation because there will be an excess of stations for the few HFCVs in the market. All of the hydrogen stations will be underutilized for a period of years. During the first several years of operation there will be fewer than 10 HFCV fills (at perhaps 5 kg each) per station per day on average. At very low demand the technology choices to provide hydrogen include truck delivery of hydrogen from current industrial hydrogen gas suppliers or from excess hydrogen production at some refineries and chemical plants directly to the filling station where it is stored, much as in today's gasoline station. Where hydrogen is not available, small skid-mounted natural gas reformers or small water electrolysis systems could be used to generate hydrogen at the refueling station. As demand grows, these small units might be moved to new areas and replaced by larger facilities. The initial hydrogen cost for all of these options is high, but as the hydrogen sales volume increases the cost will decline. The hydrogen sales volume might increase quickly during this early stage if hydrogen-powered internal combustion engine vehicles (ICEVs) as well as HFCVs were also filling at these stations. Box 3.1 contains a discussion of the hydrogen ICEV.

To illustrate this point, for a 500 kg/d natural gas reformer (about one-third the size of a full commercial-scale unit needed for a full-size refueling station) producing hydrogen at a station and dispensing at 5,000 pounds per square inch (psi), the hydrogen cost will be \$3.50/kg when operating at full capacity (70 percent). Full capacity will provide about 70 refuelings per day. For this station in the very early years when there are only 10 refuelings per day, the hydrogen cost increases to \$7.70/kg.

Midtransition Infrastructure for Expanding Hydrogen Availability to Full-size Stations

With increasing HFCV sales, hydrogen demand will eventually catch up to the initial capacity of the first stations. In addition, coverage must expand to cover distant suburban areas as well as some rural areas and highways between urban areas. The amount of hydrogen needed around any urban center would still be small, which favors hydrogen technologies that do not need a large hydrogen distribution system. This includes distributed reforming of natural gas or renewable liquid fuels such as ethanol and electrolysis of water using electricity. The distributed approach uses large appliance-type devices located at the refueling site to convert the raw material to hydrogen. These devices would be factory

BOX 3.1
The Hydrogen-powered ICEV

Hydrogen-powered ICEVs (HICEVs) have been demonstrated by several auto companies including Ford and BMW. BMW even offers a model to the public for lease. The hydrogen ICEV has many of the same advantages as an HFCV in that the fuel can be made from various sources and the primary combustion product is water. However since the HICEV still uses an oil-based lubricant, a small amount of this is consumed and very small CO_2 releases result. In addition since combustion still takes place at a high temperature, some NO_x emissions also result.

An HICEV can be about 25 percent more efficient than a gasoline ICEV, but this is offset by increases in engine cost resulting from the light nature of hydrogen. It is difficult to get enough air and hydrogen into the cylinders at normal pressure. Either much bigger cylinders are needed or boosting the air pressure through turbocharging or supercharging is needed to achieve the power a gasoline engine provides. Either choice increases the cost.

Perhaps the most difficult technical problem to resolve is storing enough hydrogen on board to travel a reasonable distance. Since the efficiency of the HICEV is only about 60 percent that of an HFCV, much more hydrogen is needed to travel the same distance. This probably will limit the number of HICEVs that enter the market.

Because of the ease in converting an existing gasoline ICEV vehicle design to an HICEV it is likely that if a hydrogen infrastructure were built, some number of HICEVs would also be built. Since the cost of this conversion is much lower than the cost of an HFCV the number could be influenced by future carbon policy or other government actions aimed at increasing hydrogen demand in order to reduce hydrogen cost especially at the beginning of the transition.

manufactured and delivered to a site ready for installation. Research and development has focused on small-scale natural gas reforming and water electrolysis.

This distributed approach likely will include adding hydrogen generating and storage to some existing gasoline filling stations as well as building retail stations specifically designed for hydrogen. The size of the stations can also be increased to become commercial scale (1,500 kg/d), reducing the cost of hydrogen further.

Full-size hydrogen stations, however, require much larger footprints than today's gasoline equivalent. Typical modern gasoline stations are approximately 6,500 square feet, including a mini-mart. Area requirement is a very significant issue because suitable large sites may be difficult to find in urban areas and acquiring the needed permits is also likely to be difficult. DOE analysis estimates that the footprint for an existing gasoline station will have to be increased by about 7,200 square feet for either a full-size natural gas reformer or a water electrolysis system (Gronich, 2007). Even a smaller unit (e.g., 100 kg/d), would require about 2,200 square feet additional area. In any case, these are significant increases that will limit the number of existing sites that could possibly be used for dispensing hydrogen. This opens up the possibility that many of the hydrogen refueling sites will be at nontraditional locations such as shopping malls and big-box retailer parking areas or even auto dealerships. Box 3.2 discusses the latter option.

Distributed natural gas reforming is the lowest-cost method of delivering hydrogen to an HFCV during this period of low but growing demand for hydrogen. Most large urban areas have an existing natural gas infrastructure allowing its use in such places. For locations in which natural gas is not available, the outer reaches of population centers, or areas between cites along highways. other methods are needed. Water electrolysis is a proven, higher-cost, method of hydrogen production. Table 3.1 summarizes hydrogen

BOX 3.2
Auto Dealers Selling HFCVs and Hydrogen

Auto dealers selling and servicing HFCVs will need hydrogen on-site for initial fills and likely for some types of service. They may be willing to either sell hydrogen to the general public or to provide land to a fuel supplier at their dealer site. It would be a selling point, especially in the early stages of the transition, because dealers could assure customers that hydrogen fuel would be available at least at their facility and neighboring dealerships. Personnel at the dealership would have to be familiar with fueling procedures anyway, so the need for specially trained personnel would be reduced. Many dealerships also would have sufficient land area, unlike most current gasoline stations.

To see if this concept is feasible, the committee examined General Motors dealerships in the Los Angeles Basin. There are 131 GM dealers in the 2,027 square mile basin, or one for every 15.5 square miles. On average, there is one GM dealership within 2.2 miles of every person, which is not a great driving distance to refuel.

Although this is only one example, and indeed a very simplified way of viewing the concept, it appears that dealerships could contribute to hydrogen availability early in the transition. Dealers and customers would have to adapt to a different mode of business, but this should be feasible. Many other types of locations would also be needed because the number of auto dealerships is small compared to the number of gasoline stations.

TABLE 3.1 Dispensed Hydrogen Costs for Distributed Generation[a]

Process	Future Technology ($/kg)	Current Technology ($/kg)
Natural gas reforming	2.60	3.30
Electrolysis	5.60	7.20

NOTE: Production plant is 1,500 kg/d; natural gas costs $6.10/MBtu; electricity costs $0.08/kWh (see Chapter 6 for other assumptions).

[a]Costs are at the dispensing nozzle and are derived from the H2A model; they represent the final dispensed hydrogen cost using either technology demonstrated today or with anticipated future improvements and a minimum of 500 installed units.

costs for the main distributed options. Reforming a bioliquid such as ethanol also is possible, although that is in an earlier stage of development (Paster, 2007a). The alternative technologies to natural gas reforming could be useful because they could lead to lower CO_2 emissions than natural gas reforming. Electricity for electrolysis could be generated by low-CO_2 methods, such as wind power, solar technologies, or nuclear methods. Reforming ethanol or another bioliquid could yield much lower CO_2 releases, depending on how the bioliquid is produced. This may become an important distinction depending on future carbon policy and regulations. For example, California currently requires that at least 33 percent of the energy used to produce hydrogen dispensed at state-funded stations be provided from renewable sources.

Late Transition to Self-sustaining Centralized Hydrogen Production

At some point in the hydrogen transition, the drive to increase production and to lower both the cost of hydrogen and the CO_2 emissions associated with hydrogen production would shift the emphasis away from distributed production toward very large centralized production plants. This is likely to start between 2025 and 2030 in the MPR case. These central plants could use a variety of primary feedstocks including natural gas, coal, and biomass. In the longer term there could be additional technology options such as hydrogen from high-temperature nuclear and concentrated solar, photobiochemical, and photoelectrochemical methods and from centralized electrolysis using solar or wind energy.

The concept of centralized hydrogen production is very different from today's centralized refining system for making gasoline, which is characterized by a few very large refining centers. They serve nearby population centers but also feed a large pipeline network for delivering to areas hundreds or thousands of miles away. Gasoline is cheaper to ship by pipeline than hydrogen (on an energy-equivalent basis). As a result, the optimal distribution system for hydrogen would lead to smaller-sized plants located closer to population centers, with few if any long-distance pipelines.

The cost of producing hydrogen at a central plant and delivering it to a station is highly dependent not only on the feedstock cost and conversion technology, but also on the size of a commercial plant and the method and distance of delivering the hydrogen. In a fully developed hydrogen economy, delivery and dispensing of hydrogen could cost as much as its production and consume significant energy.

The first central hydrogen production plants could be needed as early as 2025. With the long lead times needed to plan, permit, and construct large plants, the decisions on what type of feedstock to use would be made about 5 years prior to this. These plants will have to use commercially proven technology. Natural gas reforming and coal gasification technologies are available today. The cost of hydrogen from both feedstocks is similar when feedstock costs are $6 per million British thermal units (Btu) for natural gas and $27 per ton for coal. Twice as much CO_2 is produced in a coal-fed plant as one using natural gas. Both types could include CO_2 capture technology (at 80 to 90 percent efficiency), which produces a high-purity CO_2 stream that can then be sequestered in underground geological formations. Although CO_2 capture is a proven technology, CO_2 sequestration (i.e., permanent disposal underground) has not yet been adequately demonstrated for commercial readiness.

Biomass gasification for hydrogen production is under development but not commercially ready yet. It is likely that with continued emphasis on development it could be ready for commercial decisions in about 2025. Hydrogen from biomass gasification is more expensive than from natural gas or coal, but the net CO_2 releases are very low if land use issues can be kept small (see Chapter 4, "Alternative Technologies for Light-Duty Vehicles"). If CO_2 sequestration is not commercially available when central plants are needed, biomass could be an important source of hydrogen that would not contribute to carbon emissions.

Table 3.2 reviews the estimated costs of central hydrogen production. The costs shown in this table are derived

TABLE 3.2 Centralized Plant Gate Hydrogen Production Costs

Process	Future Technology ($/kg)	Current Technology ($/kg)
Natural gas reforming (379 tons/d)	1.50	1.60
Coal gasification (306 tons/d)	1.50	1.90
Biomass gasification (155 tons/d)	1.80	2.50

NOTE: Costs are at the plant gate and represent a learned out cost derived from the H2A model for either technology understood today or with anticipated future improvements. CO_2 capture costs are included for coal gasification but not the other technologies. CO_2 sequestration costs are not included for any technology. Future delivery and dispensing costs of about $1 to $2/kg (pipeline) or $3.50/kg (liquid) must be added to the production cost for the final delivery cost (Paster, 2007b). Feedstock costs: natural gas $6.10/MBtu; coal $27/ton; biomass $38/ton.

from the H2A model that is the basis of the resource needs analysis in Chapter 6. These costs are plant gate costs. Delivery and dispensing costs have to be added on a case-by-case basis for a final dispensed cost that could be directly comparable to those in Table 3.1 for distributed hydrogen generation.

As with distributed hydrogen production, it is unlikely that centralized hydrogen production will be done using only one feed type or method. The best method in any particular case will be determined based on regional and local issues, including resource availability and hydrogen delivery options, all within any future existing carbon policy.

Hydrogen Safety

Safety issues, both real and perceived, along with the creation of appropriate codes and standards are significant barriers to the introduction of HFCVs and the hydrogen refueling infrastructure and are a significant risk to achieving the maximum practicable penetration rate for hydrogen vehicles. Safety issues along with the need for codes and standards were addressed in *The Hydrogen Economy* (NRC, 2004) and in the FreedomCAR and fuel partnership review (NRC, 2008). The latter source noted that more urgency is needed to address these issues. It states that "work on safety, codes, and standards is an essential federal role. The individual companies and states cannot do it on their own. The manufacturers want and need uniform national (and hopefully international) standards so they can market on a worldwide basis."

HYDROGEN FEEDSTOCKS AND TECHNOLOGIES

The maximum practicable HFCV case starts with a few thousand HFCVs sold in 2012 and mass production sales (from 500,000 unit production systems) beginning several years later (after 2015). The questions about hydrogen production then are the following: (1) Will the distributed technologies be ready for commercial use about 2012 to 2015? (2) Will the centralized technologies be ready for commercial decisions about 2020 to 2025? The committee reviewed the readiness of many technologies and includes the following technologies as meeting the criteria of furthest developed and lowest cost: (1) distributed natural gas reforming, (2) distributed water electrolysis, (3) centralized coal gasification with carbon capture and sequestration (CCS), (4) centralized natural gas reforming, and (5) centralized biomass gasification. The discussion below addresses the technologies in their current state of development (current state) and their potential for further improvements (future state). Costs for the current state and future state technologies are based on DOE's H2A model, which the committee reviewed and accepted.

Distributed Natural Gas Reforming

Catalytically reforming natural gas is now the most common method of producing hydrogen at a large scale for refineries and chemical plants. The challenge for near- and midterm use in a distributed system is to package this industrial process into small units or "appliances," which can be installed inexpensively at existing retail stations. Making hydrogen directly at the filling station in a distributed system eliminates the need for a large and expensive hydrogen transportation infrastructure. This concept will be very important in the early and mid years of the transition. This concept includes all of the processes that are found in large-scale commercial plants, including the catalytic reformer, catalytic shift reactors, and hydrogen cleanup processes to achieve the appropriate hydrogen purity. Several companies are developing this concept and have demonstrated small appliances that are practical and have an appearance such that they would not look out of place in neighborhood filling stations (James et al., 2007).

The entire process has been successfully demonstrated, and most future developmental work now is directed at improving the efficiency and reducing the cost of equipment. Engineering cost analysis indicates that the overall cost of producing, storing, compressing to 5,000 psi, and delivering hydrogen to a car at a future full-scale 1,500 kg/d station is now \$3.30/kg, with the potential to reduce this further to \$2.60/kg.[1] If the vehicle stores hydrogen at 10,000 psi, as most manufacturers are proposing, the cost will be even higher, but the vehicle will be able to travel farther.

Natural gas reformers appear to offer the lowest-cost method of producing hydrogen for the near and mid term and as such receive continued industrial emphasis. The cost of hydrogen is, however, highly dependent on the size of the reformer (in kilograms per day) and on the cost of natural gas. The hydrogen cost increases with a smaller-size reformer system and with higher-cost natural gas.

To illustrate this point, the natural gas price used for estimating a \$2.60/kg cost just mentioned is \$6.10 per million Btu. If the natural gas price is increased to \$10 per million Btu then the hydrogen cost is increased to \$3.30/kg.

The ability to use distributed natural gas reformers in various sizes could be very useful during the initial transition to hydrogen to keep capital costs as low as possible—in

[1]At a natural gas cost of \$6.10/MBtu. All of the costs of making hydrogen in this chapter are based on the assumptions discussed in Chapter 6 unless noted in the text. A kilogram of hydrogen has approximately the same energy content as a gallon of gasoline and is sometimes referred to as a gasoline gallon equivalent (gge). To compare the cost of hydrogen with the cost of gasoline one must also take into account the mileage of the vehicles. In the Chapter 6 analysis, the fuel cell vehicle achieves two times the fuel economy of the gasoline ICE on a gge basis. In this analysis, an HFCV using 1 kg of hydrogen will then travel the same distance as a gasoline ICEV using 2 gallons of gasoline. Thus, hydrogen purchased at \$2.60/kg for an HFCV yields the same cost per mile as gasoline purchased for a conventional vehicle at \$1.30/gal.

essence, matching the size of the reformer to the localized demand for hydrogen. This could, however, lead to higher hydrogen costs because cost increases as reformer size decreases. For example, if the 1,500 kg/d reformer mentioned above is reduced to 500 kg/d size, then the hydrogen cost increases from $2.60/kg to $3.50/kg. Further reducing the reformer size to 100 kg/d increases the hydrogen cost to $6.20/kg.

Significant progress has been made in developing distributed natural gas reforming such that it could be ready for commercial use in the early transition years (2015 or earlier). Even with this success, still further efficiency increases and cost reductions are possible.

Distributed natural gas reforming (DNGR) will release some CO_2 into the atmosphere. It is not feasible to use CCS technologies in this distributed process to capture and sequester the CO_2 because of the large number of sites and the small size of each of them. However, on a WTW basis, an HFCV using hydrogen from DNGR units will emit less than half that emitted by a conventional ICE vehicle using gasoline, as shown in *The Hydrogen Economy* report (NRC, 2004).

The greatest potential challenges to the use of DNGR technology in the hydrogen transition are identifying and permitting land use for the refueling sites and the cost and availability of natural gas.

Distributed Water Electrolysis

Electrolysis of water is now a common way to produce small amounts of hydrogen. Scaling up an electrolyzer to full size (about 1,500 kg of hydrogen per day) has been demonstrated, and commercial alkaline electrolyzers are available in various sizes. For instance, Norsk Hydro reliably operates a 600-700 kg/d alkaline electrolyzer. Water electrolysis could play a role in the early to mid stages of the hydrogen transition because of the advanced stage of technology development, the widespread availability of electricity, and the relatively simple operation of an electrolyzer.

In alkaline electrolysis, water under an applied voltage dissociates into hydroxyl ions and hydrogen on one side of a wetted mat of sodium or potassium hydroxide solution. The hydroxyl ions traverse the wetted mat and form oxygen on the other side. Electrodes are inexpensive nickel. This is different from the polymer electrolyte membrane (PEM) process considered in many studies, including *The Hydrogen Economy* (NRC, 2004). That process involves transport of hydrogen ions across a polyelectrolyte membrane under an applied voltage using noble metal-based electrodes. The alkaline process is favored over the PEM process primarily because it does not require the costly membrane and platinum-based electrodes of the PEM process and scales up more economically at scales above 25 kg/d (Harg, 2007).

Electricity is the largest cost component of hydrogen production by electrolysis. Because the primary source and hence the cost of electricity vary considerably by region, state, and in many cases locally in the United States, the cost to make hydrogen varies widely. For example, in 2004, industrial electricity averaged 9.5 cents/kWh in California, 8 cents/kWh in Vermont, and 3.9 cents/kWh in Wyoming. This translates to the electricity cost component of making hydrogen varying from $2.10/kg hydrogen in Wyoming to $5.00/kg hydrogen in California at 74 percent electrolyzer efficiency. At a design hydrogen production rate of 1,500 kg/d and an average industrial electricity cost of 8.0 cents/kWh, the cost using current technology is $7.20/kg hydrogen. Future technical improvements are evolutionary in nature, which could result in a total hydrogen cost of $5.60/kg hydrogen (Fletcher, 2007).

The sensitivity to electricity cost is about $0.50 in hydrogen cost for each 1 cent/kWh in electricity cost, so for 2 cents lower electricity cost than the average, the hydrogen cost would be $1.00/kg lower. For the hydrogen cost from electrolysis to be in a competitive range with other feasible sources of distributed hydrogen supply (about $3/kg), electricity would have to be available at an unrealistically low cost of 2 cents/kWh (Levene, 2007). Well-to-wheels CO_2 emissions from distributed water electrolysis (DWE) could be higher than for DNGR if the electricity is generated with the current U.S. feedstock mix.

The greatest challenge to the use of DWE for hydrogen is cost. However, electrolysis of water still may be used where lower-cost methods are not available (natural gas reforming) or where environmental advantages are possible through the use of solar or wind power for electricity. In selected areas with no other source of hydrogen, policies could be put in place to help supply during a growing hydrogen transition.

Central Coal Gasification with Carbon Capture and Sequestration

Commercial large-scale high-pressure gasification plants have been used for many years to produce a syngas (carbon monoxide [CO] and hydrogen) for either power generation or further conversion to chemicals. These plants run on heavy oils, petroleum coke, or coal. In this sense the basic core technology is technically and economically ready.

Although the core gasification technology is commercially available, the overall plant design and integration of the gasification module with the downstream processing modules can still be improved through further development to lower the production cost (Litynski, 2007). Co-producing power and hydrogen at a large coal gasification facility is likely to be the lowest-cost method for making hydrogen (NRC, 2004). This concept also has not yet been demonstrated at a large scale although all of the processing modules needed are commercially available.

The largest single challenge for widespread use of coal to make hydrogen is to lower CO_2 releases to acceptably low levels. In the case of an oxygen-supplied gasification, capturing CO_2 is relatively inexpensive. This CO_2 must then be piped to a CO_2 sequestration site and injected into underground storage. In a future with constraints on carbon emissions it is likely that if coal is to be used to make large amounts of hydrogen, CO_2 sequestration must be adequately demonstrated and commercially available.

Large-scale CO_2 sequestration from an integrated coal gasification plant with CCS has not yet been adequately demonstrated. Several partnerships have been formed to further develop and demonstrate CCS technology involving DOE, 41 states, and more than 400 organizations. Seven sequestration injection tests have been awarded funding through these partnerships (Litynski, 2007). The goals of this work are to verify that the costs of CO_2 capture and sequestration are not high (less than 10 percent increase in power or hydrogen costs) and that any sequestered CO_2 will remain so (less than 1 percent leakage after 100 years). Since it will take many years to complete this program, success may not be known prior to the start of the hydrogen transition.

The cost of hydrogen produced at a central gasification plant is estimated to be \$1.50/kg at the plant gate (Joseck, 2007) for a 2,400-ton-per-day plant that makes 306,000 kg/d. This does not include costs for carbon sequestration, which should be low as a percent of the total hydrogen delivered costs (NRC, 2004)). Additional pipeline delivery and dispensing costs could increase the final delivered cost to \$2.60/kg (Paster, 2007a). Table 3.2 summarizes hydrogen plant gate production costs for the key central production technologies. The cost of making hydrogen from coal is somewhat insensitive to the price of coal, because most of the overall cost is related to the large capital cost of the plant. For instance, tripling the coal cost, from \$27 to \$81 per ton increases the hydrogen cost by about \$0.50/kg.

The greatest challenge to the use of coal gasification for hydrogen production is demonstration of the costs, capacity, safety, and risks of long-term carbon capture and sequestration. Although coal gasification is a commercially available technology, to reach the future cost estimates shown in Table 3.2, some further development is needed. Standardization of plant design, gas cooler designs, process integration, oxygen plant optimization, and acid gas removal technology shows potential for lowering costs. Other areas that can have an impact on future costs include new gasification reactor designs (entrained bed gasification) and improved gas separation (warm or hot gas separation) and purification technologies. These technologies need further R&D before they are commercially ready.

Central Natural Gas Reforming

Steam methane reforming of natural gas is the predominant method of making large amounts of hydrogen for the chemical processing industry and the refining industry. It is a well-developed and inexpensive commercial process with no real technical barriers to its expanded use to make hydrogen for transportation. The current cost of producing hydrogen at a 380-tonne-per-day plant is \$1.60/kg at the plant gate for a natural gas cost of \$6.10/MBtu (H2A model runs). This cost is similar to that of hydrogen from a coal gasification plant. Adding on delivery and dispensing costs could increase the final supply cost to about \$2.90/kg. Because steam methane reforming is such a well-developed commercial technology, little future cost improvement is expected. As shown in Table 3.2 the future plant gate cost is anticipated to decrease to \$1.50/kg at the same natural gas cost.

There are two primary issues with using significant natural gas resources to make hydrogen for transportation. One is CO_2 releases, and the other is the availability and cost of natural gas. The CO_2 associated with hydrogen production and use in an HFCV is about half of that associated with a gasoline ICEV. From this perspective, using natural gas to make hydrogen results in lower CO_2 emissions. As with coal gasification, this CO_2 can be captured and sequestered, and if done, the CO_2 releases will be very small.

The largest barrier to widespread use of central natural gas reforming technology is the availability and price of natural gas. Since the incremental natural gas supply to the United States is currently from imported liquefied natural gas (LNG) and imports are expected to grow in the future, using natural gas to make hydrogen will increase the requirement for imports. Although natural gas pricing is a complex issue, it is likely that a significant number of HFCVs using natural gas will increase natural gas demand enough to affect natural gas prices, which in turn would have a great impact on the power sector (a heavy user of natural gas). If natural gas is the source of hydrogen for 10 million HFCVs, the demand for natural gas in the United States would increase by about 2 percent, probably not enough to have a significant impact on prices. However, 50 million cars, less than 20 percent of the light-duty vehicles expected in the country in 2020, would increase natural gas demand by 10 percent, which probably would increase prices.

Centralized Biomass Gasification

Biomass may consist of waste by-products of agriculture, such as corn stover or wheat straw, or forest residues; it also can be expressly cultivated as an energy crop, for example, switchgrass. It could be gasified in the same type of high-pressure oxygen-fed gasifier used for coal, but that alternative may not be the best for gasifying biomass to make hydrogen. High-pressure gasifiers tend to be less economical at small scales than at a large scale. Much of the feedstock and some of the process issues with biomass gasification are similar to those for biofuels discussed in Chapter 4.

Because biomass is produced in relatively small quantities per farmed acre in a dispersed agricultural system, a biomass plant is likely to be much smaller than other central production methods such as a coal gasification plant. The size of any biomass plant will be determined by its proximity to a large acreage of land that has available agriculture biomass, that is appropriate to grow energy crops, or that contains forest residue, while also being near large population centers. Biomass starts out with an inherent environmental advantage versus coal or natural gas since CO_2 is cycled from the atmosphere in its growth and cycled back to the atmosphere in its processing, which can result in little net CO_2 emission to the atmosphere without the need for carbon sequestration if land use issues are kept to a minimum.

Given the economic reasons for smaller plants, a different gasification technology could prove to be a better choice for making hydrogen from biomass. There have been several small-scale (100 to 300 tons per day of biomass feed) biomass gasification test plants operated using a low-pressure, indirectly heated air gasification process, for example, a 300 tons per day Battelle gasifier unit producing fuel gas for electric power generation in Vermont. In this process the biomass is gasified with air at high temperature and atmospheric pressure to produce a mixture of CO and hydrogen. Because the Battelle gasifier operates at low pressure, it has the potential to be less capital intensive than the high-pressure oxygen gasifier, which also requires an air separation unit for oxygen feed.

DOE bases its future technology plans on the indirect-heated low-pressure air gasifier type plant at 155,000 kg hydrogen per day. The biomass feed rate required for a plant this size is 2,125 tons per day (Mann, 2007), and the land needed to supply such quantities is about 180 square miles for an energy crop such as switchgrass. The number of available sites must be determined that could support such a large plant and still have acceptable delivered biomass cost and delivered hydrogen cost with limited land use issues. The DOE estimates roughly 50 potential sites throughout the country at current biomass yields and upward of 100 sites with future crop technology (Joseck, 2007).

With future technology, DOE targets biomass crop yields per acre increasing 50 percent because of applied plant genetics and crop rotation practices. This is based on a model that assumes switchgrass, a 45 percent component of projected biomass availability, will develop in crop yield similar to the achievements with corn. However, the potential for yield improvements in other components of biomass, for example, corn stover and wheat straw, is uncertain.

The total hydrogen production cost at the plant gate based on this future technology is \$1.80/kg hydrogen with a biomass feedstock cost of \$38 per dry ton (Table 3.2). Delivery and dispensing will add additional costs of about \$1.70/kg (pipeline delivery is higher than for coal plants because biomass plants are smaller) to \$3.50/kg (liquid truck delivery) to the plant gate cost. Total projected biomass availability at \$38 per dry ton is 200 million dry tons in 2015 and 500 million dry tons in 2025, based on DOE estimates reviewed with this committee (Hess, 2007). If all of the 500 million dry tons were converted into hydrogen, this would be about 37 billion gallons of gasoline equivalent, or 26 percent of today's gasoline market.[2]

Biomass gasification is promising, but much remains to be done to put it on a solid basis. Crop yield assumptions need to be demonstrated. Preparation of the different types of biomass feedstock for ease of delivery and reliable processing in the gasifier needs to be determined. The true number of sites that can actually supply large amounts of biomass without incurring large land use problems has to be understood.

Significant research, development, and technology demonstration is required before the future costs shown in Table 3.2 can be achieved. Although individual parts of the biomass gasification process have been demonstrated, the entire process has not been demonstrated. Bench-scale, pilot plant, and semicommercial-scale work is needed to have a firm basis for scale-up to a 2,125-ton-per-day plant or larger. Also, gas cleanup and separation technologies into pure hydrogen need to be demonstrated while dealing with contaminants and tar. The committee judges that technology readiness by 2015 will be difficult to achieve; several years more may be needed. However, if successful, hydrogen supply from biomass gasification could phase in with other supply sources of hydrogen.

HYDROGEN FUEL CELL VEHICLE TECHNOLOGIES

The HFCV is an all-electric automobile. It differs from previous generations of electric cars in that the power is provided by a hydrogen fuel cell on an as-needed basis. Previous generations of electric vehicles stored energy only in batteries whereas the HFCV stores its energy primarily in a hydrogen tank. Most major automakers have demonstrated different iterations of HFCVs, with each version demonstrating new concepts to achieve performance that is approaching that of today's gasoline vehicles. Some HFCV prototype vehicles also include some degree of energy storage in batteries using the same technology that is in today's gasoline hybrid vehicles. These prototype vehicles have demonstrated significant success in overcoming difficult technological challenges, such as reducing the size and weight of the fuel cell and improving operation in cold weather. However, even with the significant improvements of the past few years,

[2]The biomass gasification technology discussed in *The Hydrogen Economy* is based on a conventional high-pressure oxygen gasifier at just 24,000 kg hydrogen per day, a small scale to ensure economic biomass supply (NRC, 2004). This is in contrast to the larger plant and broader feedstock supply reach envisioned in the future technology and the low-pressure indirect heated gasifier of the future technology discussed above. The biomass feed rate for this small plant is 442 dry tons per day from a 57 square mile collection area. The total hydrogen cost estimate is \$7.00/kg hydrogen at a biomass feedstock cost of \$53 per ton. The total cost is particularly high since hydrogen liquefaction and tanker transport are required because pipeline supply is not economic at such a small scale.

there are still several areas that need further improvement. The key component technology challenges for the HFCV are (1) making the fuel cell system as durable and cost-effective as today's gasoline internal combustion engine vehicle, and (2) engineering a small, lightweight hydrogen storage system to provide an acceptable driving range, 300 miles or more. The degree of success in these areas will determine when the HFCV can be commercialized and how effectively it can compete with today's ICEVs. This section discusses these issues and provides a context for how they can be viewed to meet the maximum practicable rate of penetration for HFCVs.

Recent History

Several major automakers have been developing the HFCV in earnest since the early 1990s. By the mid-1990s Daimler-Benz announced that it would begin production of HFCVs in about 2004. Although a small number of test vehicles has been produced, commercial production has not yet begun. In 2003 GM announced that production could begin in 2010, but more recently GM has pushed this date back to 2011 or 2012.[3] Because of the highly competitive nature of the auto industry, not every automaker has made its plans available to the public.

This situation suggests the uncertainty and risks involved in introducing a new vehicle technology to the market. One of the benefits of the FreedomCAR and Fuel Partnership (FCFP) is that all of the known areas that require further development for HFCV commercialization are included with developmental targets and dates for completion specified. This permits more realistic assessments of the state of overall development. To date, only a few hundred HFCVs have been produced, with none of the advantages of mass production. Costs in high-volume manufacturing can be estimated only roughly, because several major subsystems are still in the development stage and "tight" manufacturing estimates are not available. In the traditional process for developing a new-technology power train for commercialization, once the technology is developed, hundreds of vehicles are put into experimental stressful applications (such as police vehicles) to ensure that there are no "unknown-unknowns" that could cause premature durability problems. Following that step, small-scale volume production in the thousands can begin.

Since the HFCV will require a new fuel as well as a new power train, the automotive and fuel companies are working together to develop standards for the vehicle-fuel interaction (e.g., the purity of hydrogen required for the vehicle and the fueling protocol). Hydrogen fuel standards are currently under study by the Society of Automotive Engineers (SAE) and the International Organization for Standardization (ISO) (i.e., SAE TIR J2719 and ISO 14687-2, respectively). Once the fuel standards are developed (expected in 2008), fuel companies can then begin to build a large enough number of distribution outlets to satisfy early HFCV customers that they will be able to acquire fuel safely and conveniently.

HFCVs and Fuel Cell Technology

Progress in fuel cell development has been rapid. Vehicle fuel cells are much smaller and lighter than they were just several years ago. Fuel cell costs have decreased, while performance and durability have increased considerably. However, production-quality vehicle fuel cells are still in the development stage.

To the best of this committee's knowledge, no vehicle fuel cell system has yet met the full set of rigorous automotive specifications required for high-volume sales to customers. In some sense, these requirements must be met before an automobile company can get on the high-volume manufacturing (500,000 units) cost reduction curve. Furthermore, there is a degree of uncertainty about the manufacturing cost estimates because fuel cells have not yet been manufactured on a large scale. Thus, current cost estimates are derived from detailed engineering studies, as they were in NRC (2004, 2005, 2008), Walsh et al. (2007), and Brunner (2006). On the other hand, all major automotive manufacturers have seen enough progress that, as a group, they are spending billions of dollars to bring fuel cell vehicles to high-volume production. The main debate among the manufacturers appears to be concerned with "when," not "if."

More recently, the NRC FCFP review committee noted, "Fuel cell stack life currently limits the overall demonstrated powerplant durability to only about one fourth of what is needed to meet the performance targets set forth by the Partnership. A major reduction in stack life occurs in actual vehicle applications because of the many stops and starts and transients with vehicle operations, fuel composition, and related phenomena when compared to what is observed with the testing methods and conditions in laboratory development work. In addition, as laboratory fuel cell stack lifetimes lengthen, new failure modes are surfacing and must be better understood and resolved. One such example is platinum catalyst dissolution, which impairs long-term performance. The prompt resolution of these and new failure modes, as they are discovered, is critical to achieving 2010 and 2015 targets" (NRC, 2008, pp. 56-57).

Fuel cell costs have been reduced significantly over the past 4 or 5 years. Cost projections for high-volume (500,000 units per year) automotive fuel cell production are estimated to be \$100/kW for relatively proven technologies and \$67/kW for newer laboratory-based technologies (which may be compared with the DOE/FCFP commercialization goal for 2015 of \$30/kW). The cost of platinum is 57 percent of the fuel cell stack costs and is the greatest challenge to

[3]Larry Burns, GM vice president of R&D and Strategic Planning, said, "I don't know how many of them we'll make at the time, but we should have them in showrooms by early next decade, around 2011 or 2012. Post-2012, the goal is to ramp up production to about a million vehicles a year, worldwide" (Burns, 2007).

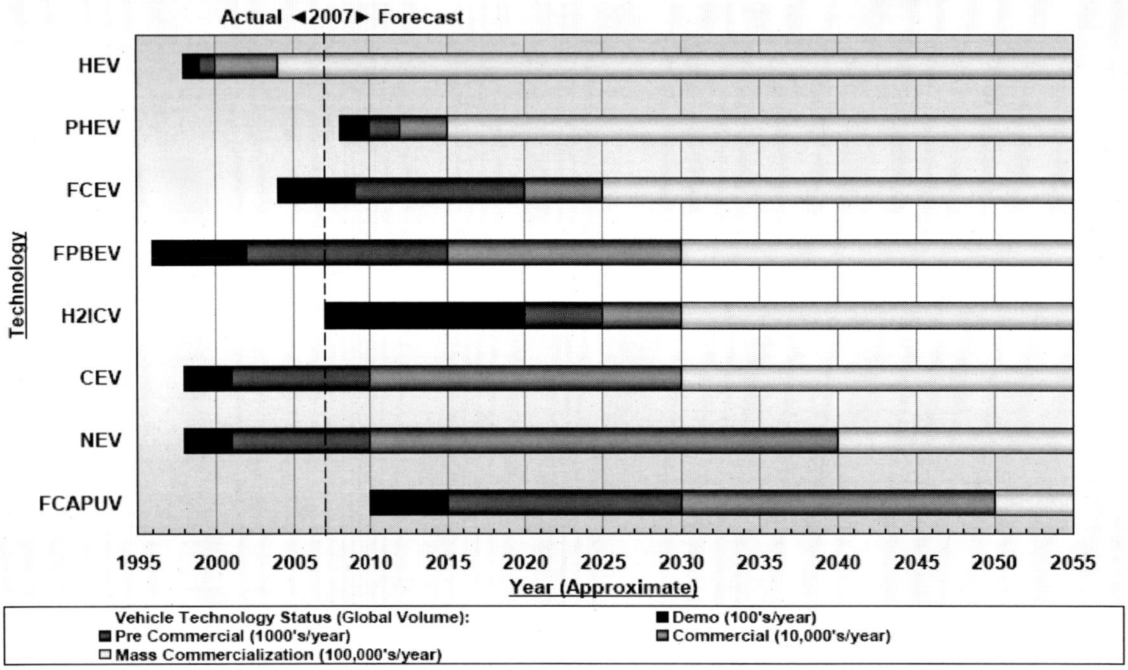

FIGURE 3.1 ZEV panel vehicle market penetration estimates. SOURCE: Walsh et al. (2007).

further cost reductions.[4] Future platinum supply is a critical issue in forward projections of fuel cell costs. Fuel cell stack life has increased to over 1,500 hours compared to the DOE/FCFP 2015 goal of 5,000 hours. Focused research on problems, together with recent advances in electrode and membrane technology, should further reduce costs and increase stack life.

Walsh et al. (2007) reports on an extensive analysis of fuel cell systems, as well as competing technologies that produce very low emission vehicles in the study called *Status and Prospects for Zero Emissions Vehicle Technology: Report of the ARB Independent Expert Panel*. These analysts visited 10 automotive manufacturers, reviewed proprietary information, and developed the following assessment after their visits and data gathering: "Each of the developers believes that the simultaneous requirements can be met but on different time schedules. For example, one major developer's objective is to compete with the 'upper' segment of ICE vehicles in the year 2020 at volumes of 100,000 units per year. Another major developer's assessment is that a commercially viable fuel cell system would be available in 2010, if a production rate of 500,000 units per year could be realized" (p. 8). The panel also noted: "There are large technical barriers that can be solved but there are other issues that are beyond the control of any single auto manufacturer. Widespread deployment of FCEVs will require continuous strong support from government agencies. This support must include a clear message of long term commitment to . . . FCEVs. These include adequate and affordable hydrogen refueling, as well as a host of sustainable financial incentives to help minimize the capitalization risks of all key stakeholders during the initial transition years. Ultimately, consumer knowledge and willingness to buy these vehicles in high volume is required" (p. 130).

Walsh et al. (2007) contained an overall estimate of market introduction time frames for the various low-emission vehicle technologies they analyzed (Figure 3.1). That ZEV expert panel's estimate is that production of thousands per year could occur by 2009, with tens of thousands per year by 2020, and then mass commercialization by 2025, with the statement that "the panel remains cautiously optimistic for fuel cell system commercialization" (p. 130). The estimates just discussed along with presentations from auto manufacturers and information included in the other resources noted were used in developing the HFCV market penetration scenarios in the Chapter 6 analysis.

The committee concludes that the current state of fuel cell development does not yet meet all of the performance and cost requirements needed for large-scale commercial production. If the recent progress in size and weight reduction, cold-weather operation, and durability improvements can be continued over the next few years, a usable fuel cell technology may be made ready for introduction by 2015. The costs of the early fuel cells are likely to be higher than the commercial targets, but these costs can drop with continued development and large-volume manufacturing.

[4]The stack is the heart of the fuel cell. It contains the membrane through which hydrogen passes to react with oxygen, generating electricity.

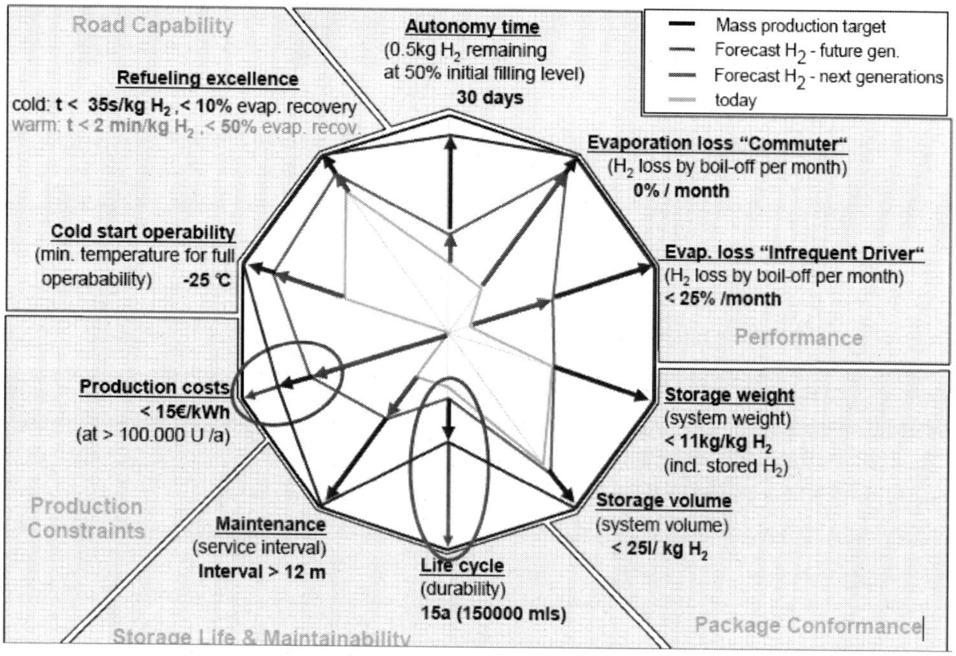

FIGURE 3.2 BMW assessment of on-board liquid hydrogen storage. SOURCE: Brunner (2006).

HFCVs and Hydrogen Storage

Hydrogen can be stored on board a vehicle as a gas, liquid, or solid. All of these techniques have been demonstrated in HFCVs in recent years. All continue to be developed through government research and industry programs. At this point there is no consensus as to which storage state will be the best long-term solution. However, there is a growing consensus that in the short term, high-pressure gaseous storage is the most practical solution.

Solid hydrogen storage systems that are made up of a low-pressure tank filled with a solid storage material with a thermal management system have the potential to be small and lightweight, which aids in overall weight savings and improved fuel mileage. They may present the best long-term storage solution but at present are still mainly in the research stage (to identify the best solid medium for storing hydrogen). Thus, it is important to develop solid hydrogen storage systems as well as to integrate them into the vehicle in an energy-efficient manner. For further discussion of solid storage technical readiness, see NRC (2008) and Walsh et al. (2007). Based on the committee's current knowledge that no solid storage medium has yet met all of the developmental targets, it is unlikely that solid hydrogen storage systems will be production ready in 2015.

Liquid hydrogen currently provides the densest form of storage, which means that the most fuel can be stored on board. BMW has recently demonstrated a liquid storage system. Two formidable problems must be overcome to make liquid storage practical for widespread use. Liquid hydrogen must be kept at about −252°C (about 20°C above absolute zero). The BMW system has impressive insulation, but some heat still gets in and gradually boils off the hydrogen. This can result in serious safety issues. The other issue is cost, currently on the order of $500/kWh, with a goal of approximately $100/kWh in the "next generation" (Brunner, 2006). The eventual goal is $15/kWh. To put these numbers in perspective, the FreedomCar targets for 2010 and 2015 are $4/kWh and $2/kWh, respectively. As shown in Figure 3.2, today's system suffers from performance, durability, and maintenance issues in addition to the noted major cost issues. The figure shows just inside the rectangle the areas in which R&D is being performed, and the goals are on the perimeter of the 10-sided figure (e.g., keeping evaporation losses to less than 25 percent per month for the infrequent driver). Inside this is the current status of progress toward the goal (e.g., less than 50 percent of the way for evaporation loss). Because of these concerns, the committee does not believe liquid storage systems will be commercially viable in the 2015-2020 time frame without unexpected breakthroughs in liquefaction and insulation.

As noted by Walsh et al. (2007), "With the exception of BMW, every other OEM [original equipment manufacturer] contacted indicated that this (compressed gas) was the only realistic short term (5-10 years) choice available and only Honda indicated that they intend to limit the storage pressure to 350 bar.[5] All the other OEMs preferred 700 bar, which will provide storage of over 50% more fuel in the

[5]One bar equals one atmosphere of pressure (14.7 pounds per square inch [psi]), so 350 bar is about 5,000 psi and 700 bar is 10,000 psi.

same space envelope and correspondingly provide almost 50% more range."

The DOE goal for hydrogen storage systems is enough fuel to travel about 300 miles (a similar range to that of today's gasoline ICEV). The amount of hydrogen needed for this depends on the fuel consumption of the HFCV. Toyota demonstrated in September 2007 a 4,145-pound, five-passenger HFCV with 700-bar compressed hydrogen storage that traveled 350 miles in real-world on-the-road conditions in a drive from Osaka to Tokyo. Toyota calculated that the vehicle is now capable of achieving a cruising distance of 466 miles. It appears that the latest HFCV designs using high-pressure hydrogen storage can meet the 300-mile goal.

Less progress has been made in meeting the cost targets for such a system. The 2005 NRC review of the FreedomCAR and Fuel Partnership listed the circa 2004 cost status as $15/kWh and $18/kWh for the 350- and 700-bar systems, respectively. The 2008 NRC review of the FCFP did not update these costs, and discussions with auto companies indicated that little has changed with regard to costs for compressed hydrogen storage.

Based on these facts, the committee concludes that compressed hydrogen storage systems that provide practical driving ranges (300 miles) should be available in 2015, but the cost will be higher than that of the current FreedomCAR targets. There is potential to lower the costs in the future through the use of lower-cost carbon fiber tanks or by using future solid storage systems.

In summary, onboard hydrogen storage to achieve a 300-mile driving range has been the greatest technical challenge of all in trying to develop an HFCV. The quest to identify solid storage materials to achieve the DOE-FCFP 2015 goals, including the cost goal of $2/kWh, is in the research stage. It is not clear at this time whether a suitable material will be identified that can meet these goals and timing targets, but to achieve the desired driving range between refueling stops, the industry is prepared to use more expensive high-pressure hydrogen storage tanks that consume more space and add to vehicle weight while research progresses toward a commercially viable solid hydrogen storage material.

Technology Basis for the Scenario Analysis

The committee concludes that not all the FreedomCar goals for 2015 are likely to be met, but the technology may be good enough for high-volume HFCVs to be introduced then anyway. For the scenarios analyzed in Chapter 6, the committee assumes that the hydrogen storage system will be larger and more costly than the targets but will be able to provide adequate driving distance. The fuel cell system will be more costly than the target initially but will provide the necessary performance expected of an early commercial vehicle. Although the initial costs will be high, there is considerable scope for continued cost improvement through technology improvements and high-volume production. For the maximum practicable case, the committee estimates that the fully learned out cost for the fuel cell drive train (the fuel cell system, hybrid battery, motor, and auxiliaries) for the automaker (OEM) will be $50/kW. This corresponds to a fuel cell system cost of $30/kW plus added costs for a hybrid battery, electric motor, and other components. Of the $30/kW fuel cell system cost, about half is due to the fuel cell stack and half to the balance of the plant. Hydrogen storage costs the OEM $10/kWh compared to DOE 2015 goal of $2/kWh for solid storage. The fuel cell cost is the same as the 2015 DOE goal, while the storage costs are higher than the DOE 2015 goal because high-pressure hydrogen gas storage was assumed in the latter.

CONCLUSIONS

CONCLUSION: If appropriate policies are adopted to accelerate the introduction of hydrogen and HFCVs, hydrogen from distributed technologies can be provided at reasonable cost to initiate the maximum practicable case. If technical targets for central production technologies are met, lower-cost hydrogen should be available to fuel HFCVs in the latter part of the time frame considered in this study. Additional policy measures are required to achieve low-carbon hydrogen production in order to significantly reduce CO_2 emissions from central coal-based plants.

CONCLUSION: Lower-cost, durable fuel cell systems for light-duty vehicles are likely to be increasingly available over the next 5-10 years and, if supported by strong government policies, commercialization and growth of HFCVs could get underway by 2015, even though all DOE targets for HFCVs may not be fully realized.

Considerable progress has been accomplished since *The Hydrogen Economy* (NRC, 2004) toward a commercially viable hydrogen fuel cell vehicle due to the concentrated efforts of private companies and governments around the world. Although considerable progress is still required in fuel cell costs, durability, and storage before commercialization can begin, the automotive industry appears committed to the technology for the long run. Thus, lower-cost, durable fuel cell systems for light-duty vehicles are likely to be available in a growing number of vehicles over the next 5-10 years, but meeting all 2015 DOE commercialization targets will be difficult.

BIBLIOGRAPHY

Bereisa, J. 2007. Energy Diversity: The Time Is Now. Presentation to the committee, June 25.

Brunner, T. 2006. BMW Clean Energy—Fuel Systems. Presented at the CARB ZEV Technology Symposium. Sacramento, California.

Burns, L. 2007. Quoted in Reuters dispatch, May 17.

Cicero, D. 2007. The U.S. DOE's Hydrogen from Coal Program. Presentation to the committee, April 19.

Duleep, K. 2007. The Hydrogen Transition and Competing Automotive Technology. Presentation to the committee, April 18.

Fletcher, K. 2007. Hydrogen Transitioning to Carbon Free Energy. Presentation to the committee, June 25.

Giove, J. 2007. FutureGen Briefing. Presentation to the committee, April 19.

Gronich, S. 2007. 2010-2025 Scenario Analysis. Presentation to the committee, February 20.

Harg, K. 2007. Hydrogen. Presentation to the committee, April 19.

Hess, R., et al. 2007. Cellulosic Biomass Feedstocks for Renewable Bioenergy. Presentation to the committee, April 18.

James, B. 2007. H2A Production Model Description. Presentation to the committee, April 19.

James, B., et al. 2007. Technology Readiness: Dist. NG and Bio-derived Liquids Reforming. Presentation to the committee, April 19.

Joseck, F. 2007. Hydrogen Program Resource Analysis. Presentation to the committee, February 21.

Kawai, T. 2007. Sustainable Mobility and the Development of Advanced Technology Vehicles. Presentation to the committee, June 26.

Lasher, S., et al. 2007. Direct Hydrogen PEMFC Manufacturing Cost Estimation for Automotive Applications. Presentation to the committee, April 25.

Levene, J. 2007. Electrolysis: Electricity Is the Key. Presentation to the committee, April 19.

Litynski, J. 2007. Summary of DOE's Carbon Sequestration Program. Presentation to the committee, April 19.

Mann, M., et al. 2007. Hydrogen from Biomass Gasification: Cost Analysis and Technology Status. Presentation to the committee, April 19.

Nicholas, M., et al. 2004. Using Geographic Information Systems to Evaluate Siting and Networks of Hydrogen Stations. *Transportation Research Record* 1880:126-134.

NRC (National Research Council). 2004. *The Hydrogen Economy: Opportunities, Costs, Barriers, and R&D Needs*. Washington, D.C.: The National Academies Press.

NRC. 2005. *Review of the Research Program of the FreedomCAR and Fuel Partnership: First Report*. Washington, D.C.: The National Academies Press.

NRC. 2008. *Review of the Research Program of the FreedomCAR and Fuel Partnership: Second Report*. Washington, D.C.: The National Academies Press.

Paster, M. 2007a. Hydrogen Delivery Infrastructure. Presentation to the committee, April 18.

Paster, M. 2007b. Hydrogen Delivery Models. Presentation to the committee, April 19.

Walsh, M., et al. 2007. *Status and Prospects for Zero Emissions Vehicle Technology: Report of the ARB Independent Expert Panel*. Prepared for State of California Air Resources Board, Sacramento, California.

4

Alternative Technologies for Light-Duty Vehicles

In keeping with its statement of task, the committee considered "whether other technologies would be less expensive or could be more quickly implemented than fuel cell technologies to achieve significant reductions in carbon dioxide emissions and oil imports." After considering a range of alternative technologies and the budget constraints of the study, the committee chose to evaluate in depth two options that received increased emphasis in the Energy Independence and Security Act of 2007 (EISA 2007): (1) evolutionary improvements in internal combustion engines (ICEs) and hybrid electric vehicles (HEVs) and (2) a biofuel option (i.e., fuel derived from biomass). Many of these vehicle technologies will be needed through 2020 to meet the significantly higher fuel economy standards required by EISA 2007. These are not the only technologies that could make a contribution to improved fuel economy and reduced carbon dioxide (CO_2) emissions, but they are the ones for which the committee felt confidence in projecting technology availability, costs, and consumer acceptance. Others are discussed briefly below.

The ultimate goal of this chapter is to estimate the extent to which continuing evolution of light-duty vehicle technologies, increased use of improved hybrid electric vehicles, and the use of biofuels can reduce oil imports and greenhouse gas emissions through 2050.

EVOLUTIONARY VEHICLE TECHNOLOGIES

Over the past 25 years, oil consumption in the light-duty vehicle fleet has grown because of an increase in the number of vehicles in the fleet and the annual miles driven, along with a shift to light-duty trucks for personal use, including sport utility vehicles (SUVs). To meet today's environmental and energy challenges, there is a need to markedly improve the fuel efficiency of the light-duty fleet in order to lower both CO_2 emissions and oil imports from their current upward trajectories.

Recent History

Before discussing evolutionary vehicle technologies, it is useful to review the evolution of some key automotive technologies since the 1960s. These changes were driven in part by two major sets of regulations:

1. *Corporate average fuel economy (CAFE)*[1] *standards.* These provisions were established by Congress in 1975. The near-term goal was to double new car fuel economy by model year 1985.
2. *The Clean Air Act of 1970.* The act (as amended in 1990) for the first time set federal limits on vehicles' emissions of so-called criteria pollutants (such as lead, particulates, carbon monoxide, volatile organic compounds, and nitrogen oxides [NO_x]).[2]

Industry responded to this pair of regulatory challenges with a "total-systems" approach to optimize the spark-ignition engine vehicle and its fuel. That approach yielded numerous important changes in major components, including switching to unleaded fuel, addition of the catalytic converter, engine computer control, port fuel injection, the four-speed automatic transmission with torque-converter lock-up, and approximately 1,000 pounds of weight reduction due to platform and material changes. Even though

[1]The Energy Policy Conservation Act, enacted into law by Congress in 1975, added Title V, "Improving Automotive Efficiency," to the Motor Vehicle Information and Cost Savings Act and established CAFE standards for passenger cars and light trucks. The act was passed in response to the 1973-1974 Arab oil embargo. An overview of these regulations is available at http://www.nhtsa.dot.gov/cars/rules/café/overview.htm.

[2]Title II, Part A of the act covers motor vehicles. It gives the administrator of the Environmental Protection Agency the duty "to prescribe (and from time to time revise) in accordance with the provisions of this section, standards applicable to the emission of any air pollutant from any class or classes of new motor vehicles or new motor vehicle engines, which in his judgment cause, or contribute to, air pollution which may reasonably be anticipated to endanger public health or welfare."

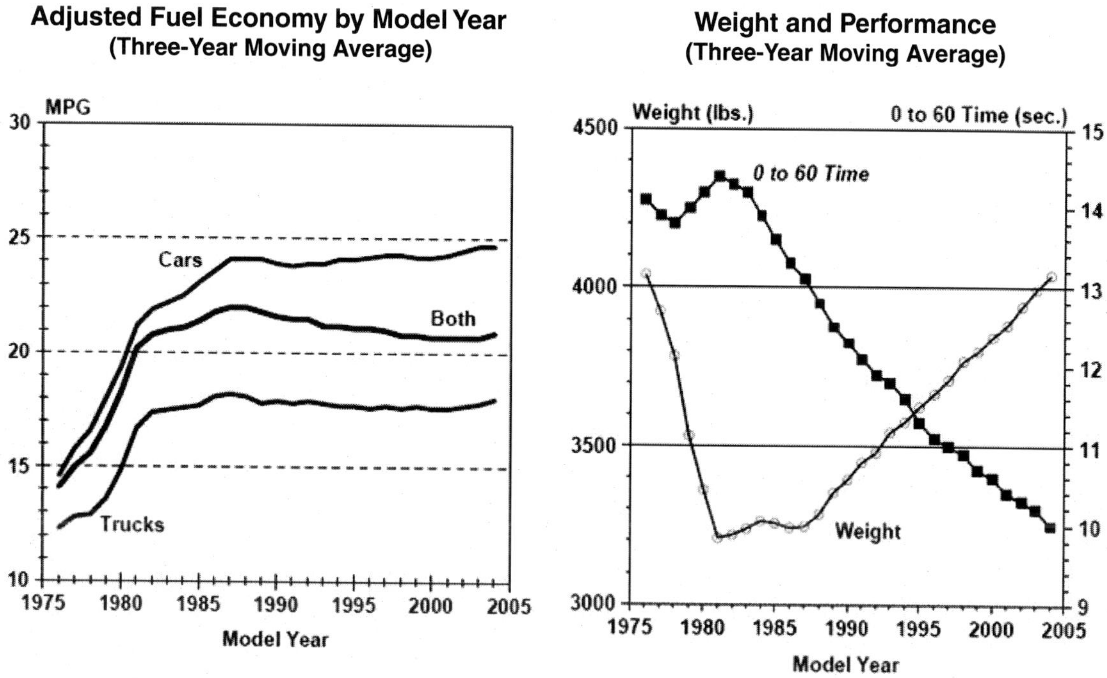

FIGURE 4.1 U.S. light-duty vehicle fuel efficiency and performance trends from 1975 to 2005. NOTE: Three-year moving average, used to smooth curves, means that for each year, what is shown is the average of that year with the two previous years. SOURCE: EPA (2006).

there were major increases in the use of plastics and aluminum, the highest percentage new material introduction was high-strength steel. Since 1980, vehicle efficiency has continued to improve even as air pollution laws and regulations have tightened, forcing vehicle designers to accommodate a multiplicity of goals. Engines, transmissions, drivetrain components, and vehicle aerodynamics have all improved remarkably, with these improvements spread among emissions reduction, improved performance, greater weight, and more power-consuming accessories in the cabin (EPA, 2005).

Since 1987, only a small fraction of these improvements have been directed to fuel economy, as shown in Figure 4.1 (EPA, 2006). After an initial marked drop in average vehicle weight and a significant fuel economy increase, most of the continuing improvement in power train technology went to overcome a steady increase in vehicle weight and to provide enhanced performance, particularly faster acceleration. The baseline case projects this trend to continue into the future because it is driven by consumer choices similar to those that have been made over the past 20 years. Although some changes in motorist attitudes can be expected, driven by higher fuel prices and an enhanced environmental consciousness, they are unlikely to produce radical improvements in light-duty vehicle fuel efficiency. However, suitable policies promoting fuel efficiency could change vehicle design priorities and result in significantly improved vehicle fuel efficiency, leading to reductions in oil imports and CO_2 emissions.

During the same period in Europe, major advances have been made in compression-ignition (diesel-fueled) engines; today, such engines form a major part of Europe's CO_2 reduction efforts. There is concern about NO_x and particulate emissions from diesel engines, and the standards for these emissions have been tightened, as have the specifications on diesel fuel to enable effective emission control technologies. This activity has resulted in the development of new after-treatment devices for NO_x and particulates for diesel engines. These technologies continue to improve.

In 1997, the hybrid electric vehicle was introduced in Japan and, in 2000, imported to the United States. In 2006, 364,845 HEVs (with 254,545 in the United States) were sold worldwide out of 68,727,429 total global vehicle sales. The National Research Council (NRC) report on the hydrogen economy and fuel cell vehicle, which was released in February 2004 but had access only to actual year-end 2002 HEV sales numbers, projected that 2006 HEV sales in the United States would represent 2 percent of the market and would increase 1 percent annually thereafter for the next 9 years and 5 percent per year for the next 10 years (NRC, 2004). The actual 2006 U.S. sales were 1.5 percent. As shown in Figure 4.2, even though gasoline prices have fluctuated dramatically from 2004 to the present, the actual increase in sales was less in 2006 than in 2005 (DOE, 2007).

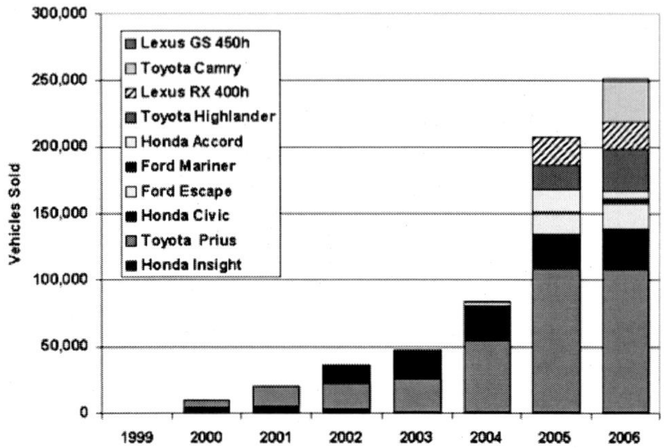

FIGURE 4.2 U.S. hybrid electric vehicle sales through 2006. Note that the order of vehicles in the key matches the order indicated in the bars of the graph from top to bottom. SOURCE: DOE (2007). Available at http://www1.eere.energy.gov/vehiclesandfuels/facts/2007_fcvt_fotw462.html.

Potential for Reducing Oil Use in Conventional Light-Duty Vehicles

This section examines options for improving fuel economy in conventional gasoline, diesel, and hybrid power train vehicles. Although not discussed explicitly in this chapter, CO_2 reductions directly follow improvements in fuel efficiency in these types of vehicles. CO_2 estimates are presented in Chapter 6. Various transportation demand reduction efforts are under way, but are not considered in this report.

Gasoline and diesel power train vehicle technologies are considered mature, but they will continue to advance and improve for the foreseeable future. The following material reviews the areas in which improvements can be expected to continue, evaluates the potential extent of these improvements, and projects their impact on fuel economy. Potential cost increases are also estimated, although these are more qualitative. For improved vehicle technology to have a significant effect on CO_2 emissions of the light-duty fleet and on oil imports, it must be directed to fuel efficiency improvements and must be included in a large fraction of all vehicles manufactured and sold in a given year; then it, or further improved versions, must continue to be manufactured for the time required to turn over a majority of the vehicles in the existing fleet. This is a decadal process.

Technical Improvements in Gasoline and Diesel-Powered Light-duty Vehicles

The approaches to improving the fuel economy for gasoline and diesel vehicles are well understood. Each of these areas offers considerable potential fuel efficiency benefits. The methods considered here include the following:

- Efficiency improvements
- Transmission evolution
- Vehicle weight reduction
- Aerodynamic improvements, reduced rolling resistance

Efficiency Improvements—Spark-ignition Engine. Technical improvements that can be applied to spark-ignited internal combustion engines include the following:

- Variable valve timing (VVT) and variable valve lift offer improvements in part-load engine efficiency. Engines only operate at full load during hard acceleration and hill climbing; the remainder of the time, engine operation is at part load, when much less power or a smaller engine is all that is needed. VVT and lift allow the engine to supply the reduced power requirement with improved fuel efficiency. The next major change is camless valve actuation (CVA), which can vary lift and timing and also allow strategies such as cylinder deactivation at light loads while simultaneously reducing valve-train friction.
- Cylinder deactivation (also called cylinder cut-out) can also reduce fuel consumption under part-load vehicle operation. Efficiency is improved by having fewer cylinders working at higher load. This technology is already being implemented, especially in V8 engines, but constitutes a small fraction of the market to date. Application to four- and six-cylinder engines is more difficult because of noise and vibration problems.
- Gasoline direct injection (GDI) and GDI combined with turbocharging allow higher compression ratio operation because of the cooling effect of in-cylinder fuel evaporation, which protects against knock. Raising the compression ratio increases efficiency (which is why diesels are relatively efficient), but conventional spark-ignition engines are susceptible to pre-ignition (knock) if the compression is too high. With GDI, the cooling effect counters the heat from the high compression. This technology benefits from variable-geometry turbines for turbo boosting and variable compression ratio. This produces more horsepower from a smaller engine, allowing weight reduction and fuel savings. This technology option is in limited production in Europe and Japan and is on a few models in the United States.
- Homogeneous-charge compression ignition (HCCI) enabled by CVA might decrease fuel consumption more than GDI with turbocharging. HCCI involves the introduction of a homogeneous air-fuel mixture, where the fuel is a gasoline-range hydrocarbon, into the cylinder. HCCI uses the same kind of "charge" as a spark-ignition engine but with higher compression, whereas a classical compression-ignition engine uses a stratified charge with a higher-boiling fuel that is injected directly into highly compressed air in the cylinder. Although HCCI has typically been considered a separate technology, its components will most likely be implemented as engine technology advances providing additional reductions in fuel consumption.

• Other changes that would result in improved fuel consumption include reduced friction in the engine and drivetrain via new materials or better lubricants; intelligent start-stop with enhanced starter and battery, which involves engine shutoff when the vehicle stops (e.g., at a red light); variable compression ratio technology; improved lean-NO_x catalyst technology; and improved engine controls and integration.

Normally aspirated spark-ignition engine efficiency is expected to improve significantly as these technologies are successfully applied. Collectively, their application can be expected to increase peak engine efficiency by about 7 to 9 percent by 2030. For turbocharged spark-ignition engines, the efficiency gain could be better than that projected for the normally aspirated spark-ignition engine by 2030 (Kasseris and Heywood, 2007). Turbocharged gasoline direct-injection engines with variable valve lift and timing could approach the efficiency of today's diesel engines (see discussion below), which will also see efficiency improvements during this time. Turbocharging is used commercially in a small fraction of light-duty vehicles today, mainly to enhance performance. These engine efficiency gains can be translated into reductions in fuel consumption and CO_2 emissions.

Efficiency Improvements—Compression-ignition (Diesel) Engines. Significant improvements to diesel engines have been made over the past 20 years, but further gains can be expected. The several technologies that will add to this include improved fuel injection, higher boost levels, series turbocharging, more advanced engine control technology, and variable valve train control. Although the major issue with diesel—emissions—has largely been resolved, current emission control technology to meet U.S. emission laws is complex and costly. Several currently produced European diesels have met the Environmental Protection Agency's (EPA's) Tier 2 and California's LEV (low emission vehicle) 2 emissions standards, although doing so subtracts from their overall efficiency. The diesel's fuel consumption is 18 to 25 percent better (lower) than that of a normally aspirated spark-ignition engine, and it is robust across all driving conditions. Diesel engines offer tremendous low-end torque and are well suited to towing and cargo hauling. With the application of the technologies noted above, diesel engine fuel consumption can be expected to be reduced further by about 30 percent over typical diesel fuel consumption today after subtracting for emissions control (Kasseris and Heywood, 2007).

Transmission Evolution. Transmission type and operation also have a significant effect on fuel consumption. U.S. drivers have favored automatic transmissions, which have been less efficient than manual transmissions. New transmission technologies provide the user-friendliness of automatic transmissions with the efficiency of a manual transmission. Developments, such as the conventional 6/7/8 speed automatic transmission and the automated manual transmission, which will soon be offered on some European models, will be almost as efficient as manual transmissions. With these technologies, transmission efficiency should be about 93 percent, a significant gain over typical automatic transmissions today (Kasseris and Heywood, 2007). In addition, transmissions with more speeds allow the engine to operate closer to its optimum efficiency, reducing fuel consumption. Continuously variable transmissions (CVTs) allow the vehicle to vary the gear ratio continuously as it goes through its driving cycle for even greater power train efficiency (Nishigaya et al., 2001; Burke et al., 2003).

Vehicle Weight Reduction. Reducing vehicle weight is a significant factor in improving vehicle fuel economy. Weight reduction reduces the energy required to move the vehicle. In doing so, the power plant may be reduced in size and weight itself, further improving fuel efficiency. Weight can be reduced by using advanced materials such as advanced, high-strength steel, aluminum, magnesium, carbon fiber composites, and plastics or by redesign to reduce the weight without using advanced materials. Estimates of potential weight reductions range from 10 to 33 percent (Weiss et al., 2000; An et al., 2001). Although weight reductions in the upper end of this range are technically possible, reductions in the range of 5 to 10 percent by 2025 are probably achievable at a reasonable cost (Duleep, 2007), although some automakers are already planning to go further.

Other Improvements. Vehicle aerodynamics, as measured by a reduction in the coefficient of drag (C_d), has improved recently by at least 1 percent per year (Hucho, 1998; Weiss et al., 2000; An et al., 2001). If this trend continued for the next 25 years, it would provide a significant reduction in fuel consumption if car size remains the same. Reduction in car size (frontal area) would result in added fuel consumption benefits because total resistance would be reduced. Vehicle rolling resistance has historically undergone an estimated reduction of 1.1 to 1.6 percent annually. An NRC committee recently concluded (NRC, 2006) that a 10 percent reduction in rolling friction is possible for today's replacement tires, and further improvements are possible.

Summary of Evolutionary Conventional Vehicle Improvements

A summary of the reductions in fuel consumption for *spark-ignition powered* vehicles from the technology advances reviewed above for the periods 2006-2015 and 2016-2025 is given in Table 4.1, based on Duleep's work (Duleep, 2007). These technology-driven fuel-consumption improvements must be focused on delivering efficiency gains (versus providing more performance or motorist comforts) and have to be broadly incorporated into the light-duty fleet if they are to have the desired impact on reducing fuel con-

TABLE 4.1 Potential Percentage Reductions in Fuel Consumption (gallons per mile) for Spark-ignition Vehicles Expected from Advances in Conventional Vehicle Technology by Category, Projected to 2025

	2006-2015	2016-2025
Engine and transmission	12-16%	18-22%
Weight, drag, and tire loss reduction	6-9%	10-13%
Accessories	2-3%	3-4%
Idle stop	3-4%	3-4%

NOTE: Values for 2016-2025 include those of 2006-2015. SOURCE: Duleep (2007).

sumption and CO_2 emissions. Combining the projections for improvements in engine, transmission, weight, parasitic loss (including friction losses, rolling resistance, and air drag), accessories, and idle-stop components suggests reductions in fuel consumption, relative to today's vehicle and performance, of 21 to 29 percent (average = 25 percent) by 2015 and 31 to 37 percent (average = 34 percent) by 2025.

Heywood and colleagues at Massachusetts Institute of Technology (MIT) have carried out a series of studies focused on light-duty vehicle fuel efficiency (Heywood, 2007; Kasseris and Heywood, 2007; Kromer and Heywood, 2007). Their studies rely on many of the technologies discussed above but include more aggressive reductions in weight and rolling resistance and aerodynamic drag. Figure 4.3 summarizes the projected fuel consumption improvements for combinations of technology for gasoline- and diesel-powered light-duty vehicles and hybrid electric vehicles

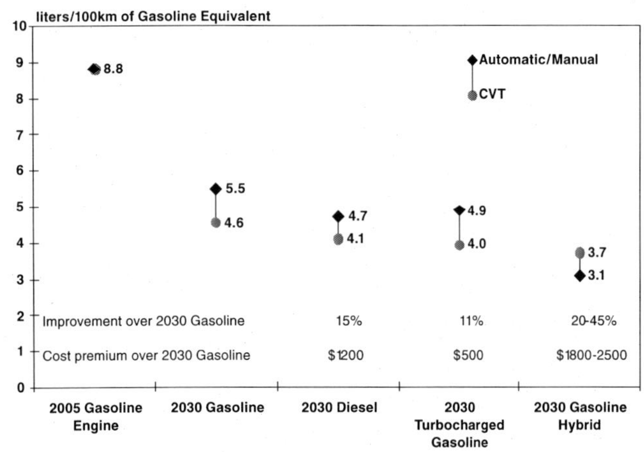

FIGURE 4.3 Fuel consumption of light-duty vehicles with different power trains using projected 2030 technology compared to a typical 2005 gasoline-powered vehicle. NOTE: To convert to gallons per 100 miles, multiply liters per 100 km by 0.426. SOURCE: Heywood (2007).

in 2030 compared to a conventional spark-ignition gasoline engine in a 2005 midsized sedan from Heywood's work.

The Heywood fuel consumption improvements result from changes in the engine and transmission and include appropriate vehicle weight reductions as well. It is assumed that the improvements are entirely dedicated to reduced fuel consumption. The 2030 *spark-ignition gasoline-powered vehicle*, with automatic manual transmission and weight reduction included, is projected to have a 38 percent reduction in fuel consumption from the 2005 version of the gasoline vehicle technology. Spark-ignition gasoline-powered vehicles with turbocharger, automatic manual transmission, weight reduction, and parasitic-loss reductions are projected to be capable of a 44 percent reduction in fuel consumption versus today's spark-ignition gasoline vehicle and an 11 percent improvement over the 2030 spark-ignition gasoline vehicle. Heywood also concludes that CVT technology could reduce fuel consumption further, reaching a 48 percent reduction in fuel consumption compared to today. The turbocharged version of the gasoline vehicle would have an estimated $500 cost premium compared to the 2030 gasoline vehicle.

Also shown in Figure 4.3 is the projected fuel consumption of a 2030 compression-ignition diesel-powered vehicle. The 2030 diesel-powered vehicle, with weight reduction, is projected to have 47 to 50 percent lower fuel consumption than the 2005 spark-ignition gasoline vehicle on a gasoline energy-equivalent basis. It also has a projected 15 percent lower fuel consumption than the 2030 spark-ignition gasoline vehicle on a gasoline energy-equivalent basis. On a fuel volume basis (kilometers per liter of diesel fuel), the 2030 diesel engine shows an additional 15 percent benefit over the 2030 gasoline engine because of the greater energy density of diesel fuel. The diesel-powered vehicle is projected to have a cost premium of about $1,200 over the 2030 gasoline-powered vehicle. If diesels achieved significant market penetration, they could increase the potential reduction in fuel consumption to 3 percent per year. However, the committee was too uncertain of the cost penalty for diesels that meet emission regulations to be willing to include increased penetration in the scenarios of Chapter 6.

Based on the work reviewed above, and assuming a baseline of 25 mpg (miles per gallon) and fuel increases starting in 2010, Duleep's projections correspond to a 2.3 to 2.9 percent compounding annual reduction in fuel consumption through 2025, or an average of 2.6 percent per year. This results in the average value from Duleep's analysis, a 34 percent fuel consumption reduction, being applied by 2025. This estimate assumes the diesel improvements discussed above, but the diesel percentage of the fleet remains constant over time. While the data from Heywood suggest that this pace could continue through 2030, the committee judges, based on the discussion in the recent NRC review of the FreedomCar Fuel Partnership (NRC, 2008), that continued weight reductions beyond 2025 would be slower to enter the fleet, delaying Heywood's potential for gasoline vehicles by about 5 years and slowing the pace of vehicle fuel consumption reduction to about 1.7 percent per year between 2025 and 2035. This leads to a 44 percent reduction in fuel consumption compared to 2009, which is Heywood's projection for the turbocharged spark-ignition gasoline vehicle with automatic manual transmission.

Further, a 0.5 percent per year reduction in fuel consumption was assumed from 2035 to 2050. At this pace, a 48 percent reduction in fuel consumption would be achieved by 2050. This improvement could be reached if Heywood's more aggressive projections for turbocharged spark-ignition gasoline vehicles with continuously variable transmissions are realized. Alternatively, the fuel consumption reductions shown in Figure 4.3 do not define the limit to how far conventional fuel economy can go. Heywood's analysis looks at a 20 percent weight reduction, relying mainly on higher-strength metals, but future carbon composites could lead to weight reductions of one-third or more. Even limited use of carbon composites to achieve a 25 percent total weight reduction would boost Heywood's spark-ignition vehicles to a 46-48 percent reduction in fuel consumption, up from 44 percent. Weight reductions through changes in the fleet mix, rather than changes in materials would have similar effects. A 10 percentage point shift from trucks to cars (i.e., increasing car share to 60 percent) would increase Heywood's spark-ignition vehicles to a 45-48 percent reduction in fuel consumption. In conclusion, the spark-ignition internal combustion engine vehicle can produce marked reductions in fuel consumption and CO_2 emissions in the intermediate and long term at little additional cost. Policy initiatives may be necessary to help speed additional weight reduction and associated fuel consumption reductions.

Thus, the committee judges that evolutionary vehicle technologies could, if focused on vehicle efficiency, reduce fuel consumption by 2.6 percent per year through 2025, 1.7 percent per year in the 2025-2035 time frame, and 0.5 percent per year between 2035 and 2050.

Increased penetration of diesel engines into the light-duty vehicle fleet was limited in the projections made here (and in Case 2 in Chapter 6) because of the poor public perception of diesels in the United States and concerns over meeting future tailpipe emission specifications. However, advanced diesel power trains could offer an additional 15 percent reduction in fuel consumption and a similar level of CO_2 emissions reductions over advanced conventional spark-ignition power trains without turbochargers and have cost advantages over hybrid electric vehicles (Kromer and Heywood, 2007). In a high-fuel-cost environment, diesels could become a growing fraction of the light-duty vehicle fleet promoted by some shifts in government positions on diesels and a positive public relations program.

Potential for Reducing Oil Use in Hybrid Electric Light-duty Vehicles

Hybrid electric vehicles and plug-in hybrid electric vehicles (PHEVs) provide increased energy efficiency by mating a battery-powered electric motor with a conventional gasoline engine. The engine can be smaller than in conventional vehicles, and therefore more economical, but acceleration can be just as great because additional power is supplied by the electric motor. The batteries can be charged in part by regenerative braking, which captures the energy that otherwise would be lost in braking. In addition, the electric motor allows the engine to be operated at or near its optimum efficiency a greater fraction of the time.

The two hybrids are similar, except that future PHEVs would have much larger battery packs that are anticipated to be able to operate the vehicle for as much as 40 miles without the gasoline engine. When the battery charge is low, the engine kicks in and operates normally, recharging the batteries and providing power. The batteries for PHEVs can also be charged from the electric grid. Thus, for some driving patterns, no gasoline at all would be needed. PHEVs are not available commercially yet because the advanced batteries that would make them viable are not ready. As battery technology improves and costs decrease, both of these hybrid power trains will become more effective and have a smaller cost differential relative to conventional engine vehicles.

All-electric vehicles would use no gasoline, but they are even more dependent on improved battery technology than PHEVs. All-electric vehicles will have CO_2 emissions that are determined by the emitted CO_2 content of the electricity on the power grid.

Hybrid Electric Vehicles

When HEVs were first developed, the technology was dedicated almost exclusively to improving fuel economy. However, as with other power trains, as HEV technology has begun to mature, the benefits of hybrid electric vehicles have shifted to now represent a blend of fuel efficiency, performance, and larger vehicles. Still, HEVs presently are the most efficient light-duty vehicles. Continued improvement relative to conventional vehicle technology is likely to occur. These improvements are expected to arise largely from improved vehicle integration, allowing tighter, more optimized control of the engine operating points. Due in part to scale economies and in part to significant reductions in the cost of high-power batteries, the incremental cost of HEVs relative to conventional technology vehicles is expected to decrease. Furthermore, HEVs will benefit from continued technological development in terms of vehicle weight and accessory loads, as well as improvements in engine technology. CO_2 emissions from hybrid electric vehicles are reduced as fuel consumption decreases, and current models, such as the Ford Escape Hybrid, the Honda Civic Hybrid, Nissan

TABLE 4.2 Comparison of Projected Improvements in Vehicle Fuel Consumption from Advances in Conventional Vehicle Technology

		From Kromer and Heywood (2007)		From Weiss et al. (2000)	
	Fuel Consumption (liters/100 km)	Relative to 2005 Gasoline	Relative to 2030 Gasoline	Relative to 2005 Gasoline	Relative to 2030 Gasoline
2005 Gasoline ICE	8.8	1			
2005 Diesel	7.4	0.84			
2005 Turbo	7.9	0.9			
2005 Hybrid	5.7	0.65			
2030 Gasoline	5.5	0.63	1.00		
2030 Diesel	4.7	0.53	0.85	0.61	1.00
2030 Turbo	4.9	0.56	0.89	0.45	0.77
2030 Hybrid	3.1	0.35	0.56	0.54	0.88
2030 Plug-In	1.9	0.21	0.34	0.38	0.615

Altima Hybrid, the two-wheel drive Chevrolet Tahoe Hybrid, and the Toyota Camry and Highlander hybrids, achieve fuel consumption reductions of 26 to 34 percent, with a simple average of about 29 percent.

As indicated in Figure 4.3, analysis from MIT indicates that CO_2 emissions and fuel consumption could be as much as 37 percent lower than those from a competitive 2030 spark-ignition gasoline vehicle (Kromer and Heywood, 2007). Table 4.2 compares the results of Kromer and Heywood (2007) and those of an earlier study, "On The Road in 2020," by Weiss et al. (2000).[3] The Kromer and Heywood (2007) study projects that HEVs will improve at a rate significantly faster than conventional gasoline vehicles. This may be expected in the earlier years because they are relatively new and should enjoy a steeper learning curve, but hybrids should be a mainstream technology before 2020.

While the committee acknowledges the significant potential for hybrids outlined in Kromer and Heywood, it concluded that most advances in hybrid technology are likely to lower the cost of battery packs (which will increase their appeal to consumers relative to conventional vehicles, and thus their market share) rather than increase fuel economy. To simplify the analysis in this report, the committee assumed that hybrids achieve a constant 29 percent fuel consumption reduction compared to conventional vehicles (which also improve each year) in each year. While this is conservative compared to Kromer and Heywood, it still leads to a 60 mpg average for new spark-ignition hybrids by 2050.

Thus, the committee judges that hybrid electric vehicles could, if focused on vehicle efficiency, consistently provide a 29 percent reduction in fuel consumption relative to comparable evolutionary ICEVs. This assumes that hybrids will have reached their peak in terms of fuel consumption reductions relative to conventional vehicles by 2009 and future improvements in hybrid fuel economy will be due, primarily, to the same technologies used to improve conventional vehicles. As a result this analysis assumes that hybrid vehicles reduce fuel consumption by 2.6 percent per year from 2010 through 2025, 1.7 percent per year in 2025-2035, and 0.5 percent per year between 2035 and 2050, the same as for evolutionary ICEVs.

Plug-in Hybrids

PHEVs are designed to be charged from the power grid and to have a significant range without requiring operation of the internal combustion engine or only using the engine under higher-power driving conditions. This range is dependent on the availability of economic, high-energy battery technology. Cost-effective battery technology for even a 20-mile trip is not available today. PHEVs could become commercially available within a few years if battery performance and costs continue to improve rapidly.

PHEVs will be significantly more expensive than HEVs. For a 30-mile-range plug-in hybrid, which is not considered viable with existing battery technology, the incremental cost is estimated to be between \$3,800 and \$4,300 over an HEV, the largest component of which is the battery, estimated to be about \$2,500 (Kromer and Heywood, 2007).

PHEVs could save more oil than HEVs. Table 4.2 shows an improvement of 1.2 liters per 100 km (about one-half gallon per 100 miles), but the actual savings would be highly dependent on driving patterns and the frequency with which

[3]Weiss et al. (2000) did not consider plug-in hybrids. Otherwise, the results of these two studies agree reasonably well.

the vehicle is plugged in.[4] However, the plug-in hybrid's CO_2 emissions reduction will not be nearly as large as suggested by their lower fuel consumption because of the CO_2 emissions associated with power generation. Today about two-thirds of our electricity is generated from fossil fuels (mainly coal and natural gas, none of which employs carbon capture and sequestering because of the expense). Still, PHEVs offer a potential path for incremental CO_2 emissions savings through increased decarbonization of the power sector—an opportunity that does not exist for HEVs.

The committee chose not to include PHEVs in the alternative vehicle case even though they have significant long-term potential. The main issue for the committee was predicting the rate of battery advancement to achieve significant driving distances. This, in the committee's view, and as presented in the recent FreedomCar Fuel Partnership review, is more than evolutionary technology (NRC, 2008). In addition, consumers' acceptance of the additional cost relative to HEVs and their willingness and ability to plug the vehicle in essentially every day are uncertain.

Conclusion on Evolutionary Technologies

CONCLUSION: Continued advancements in conventional vehicles offer significant potential to reduce oil use and CO_2 emissions through improved fuel economy, but policy measures and/or significant long-term increases in fuel costs probably will be required to realize these potential fuel economy gains in a significant number of on-road vehicles.

IMPACT OF BIOFUELS

Overview

By definition, a biofuel is any fuel that is produced from plant- or animal-based materials, generically referred to as biomass. The biofuel that is receiving the most attention today is ethanol, which can be produced from the starch and sugar in grains using fermentation and is referred to as grain ethanol. More than 6 billion gallons of grain ethanol were produced for vehicle fuel in the United States, primarily from corn, in 2007 (RFA, 2008). Ethanol can also be produced from the cellulosic material in plants using biotechnology and is referred to as cellulosic ethanol. Cellulosic ethanol production technology is evolving today and is not yet fully commercial. Higher alcohols, such as butanol, which can use the gasoline infrastructure directly, can also be produced from biomass by biological routes. Fuels that can substitute for diesel fuel, referred to as biodiesel, can be produced from plant oils and animal fats by esterification. A range of fuels can also be made from biomass using thermochemical conversion of the biomass to syngas followed by catalytic synthesis of a range of products, including gasoline and diesel. These are also biofuels because they are derived from biomass.

Biofuels provide the opportunity to sustainably produce liquid fuels for the U.S. transportation system, reduce oil imports, and reduce carbon dioxide emissions from the transportation sector because the CO_2 emitted from their combustion is captured in the next plant growth cycle. The true impact of biofuels on CO_2 emissions and oil use requires a full life-cycle analysis because fossil fuels, including oil, are used in growing and collecting biomass and, in the case of grain-based ethanol, in processing biomass into biofuels. Also, CO_2 and other greenhouse gases can be released if land use and land management practices change in conjunction with the production of biomass for energy. When land use changes result in a reduction in soil carbon, that carbon is released and emitted as CO_2. This is a major concern. Proper land use changes could result in storage of more carbon in the soil, decreasing net CO_2 emissions. Biomass-based biofuels are an emerging area that is expected to see much change in the next several decades.

This section first discusses biomass availability and some of the associated issues. It then reviews the production technology for those biofuels that are commercial or close to commercialization. It also briefly discusses biofuel technologies that have longer-term potential but are today in the research stage. The primary biofuels for transportation considered here include ethanol from corn grain, ethanol from cellulose, biobutanol from starch and sugars, and biodiesel from plant oils. At the end of the biofuels section, gasification of biomass to produce synthesis gas is discussed briefly; this can be converted into diesel and gasoline. It represents a parallel route to biofuel production. Today, it is not clear which route to biofuels (biochemical or thermochemical) will be the most economic.

An important biofuel driver in the United States is political. The Energy Policy Act of 2005 mandates 7.5 billion gallons per year of ethanol and/or biodiesel in the nation's fuel supply by 2012. The President's State of the Union Address in January 2007 increased the target volume of renewable and alternative fuels to 35 billion gallons per year by 2017. (Then in December 2007, EISA 2007 was enacted with a goal of 36 billion gallons per year by 2020.) These initiatives are acting to drive production and demand for biofuels, particularly ethanol, in the United States. Note that the energy content of ethanol is approximately two-thirds that of gasoline—thus, 35 billion gallons replaces about 24 billion gallons of gasoline. However, if engines can be optimized to operate on ethanol instead of gasoline (e.g., a higher compression is possible with ethanol), they will be more efficient, somewhat compensating for the lower energy content of the fuel.

[4] A 40-mile PHEV driven 40 miles every day and plugged in every night would use essentially no gasoline at all, saving as much as 365 gallons per year relative to a 40 mpg HEV. The same vehicle driven 20 miles every day or 80 miles every other day would save exactly half that.

Biomass Availability

Biomass is anything that grows, and since it is renewable if grown responsibly, it represents a sustainable source of feedstocks for energy, fuels, and petrochemical production. For the *present time*, easily processed agricultural crops, such as corn, will likely be the major biomass feedstock to the emerging biofuels industry. Low- or even negative-cost materials such as waste greases and cooking oils will also be used to the extent available. In the *mid term*, agricultural crop residues, forestry residues, and short-rotation trees will likely provide the feedstocks to an expanding biofuels industry. However, in the *longer term*, energy crops including perennial grasses and algae will probably have to be added to the mix to supply the amount of feedstock needed to meet national bioenergy goals.

Current annual U.S. grain production includes about 11 billion bushels of corn, 2.9 billion bushels of soybeans, 2.0 billion bushels of wheat, and significantly lesser quantities of sorghum, rapeseed, oats, and barley (Perlack et al., 2005). Higher corn prices and increased interest in ethanol resulted in an increase in corn acreage from 78 million acres in 2006 to 93 million acres in 2007 (Achenbach, 2007). Much of this increased corn acreage came at the expense of other grains, mainly soybeans. However, most of the grain grown in the United States has traditionally been dedicated to food and livestock feed supply and as such has high societal economic value. The result is that a maximum of about 25 percent of any grain is likely to be available for production of transportation fuel. This observation is based on recent experience in which corn prices rose sharply when ~20 percent of the grain was diverted to ethanol production. The U.S. Department of Agriculture (USDA) estimates that in the mid term, the United States can meet food, feed, and export demand for corn and not seriously disrupt markets with less than 20 percent of the corn crop being diverted to ethanol (USDA-OCE, 2005).

In comparison, the amount of non-grain biomass in the United States is potentially significantly larger. Walsh et al. (2000) estimated the amount of biomass currently available in the United States at less than $50 per dry ton (in 1995 dollars) to include 135 million dry tons per year of forest and mill residues and 150 million dry tons per year of agricultural crop residues. However, most of the mill residues are already being used for heat and power in the lumber, pulp, and paper industry. The total estimated biomass available at this price in the United States increases to 510 million dry tons per year when municipal wood wastes (37 million dry tons) and dedicated energy crops (188 million dry tons) are included. This is an estimate of the biomass that was projected to be economically available in 2000 for less than $50 per dry ton in 1995 dollars (Walsh et al., 2000). This would be about $68 per dry ton in 2007 dollars. The cost of biomass supplied to an ethanol production plant would have to be less than about $65 per ton to be economical with future production technologies. However, there are significant uncertainties as to how much biomass would actually be available at this price and whether it could be converted into fuel at a close-to-acceptable cost.

More recently, Milbrandt (2005) estimated the current amount of biomass that is technically available (meaning without considering costs) in the United States to be 466 dry tons (423 million dry tonnes) per year using a state-by-state geographic information system (GIS) buildup. The major components are 184 million dry tons (132 million dry tonnes) per year of forest, mill, and urban wood residues; 173 million dry tons (157 million dry tonnes) per year of crop residues (estimated using a 35 percent recovery of total crop biomass residue); and 99 million dry tons (84 million dry tonnes) per year of switch grass on Conservation Reserve Program (CRP) lands. Perlack et al. (2005) estimated that the technically available sustainable annual forest resource was about 368 million dry tons per year, of which 142 million dry tons was currently being used, leaving about 226 million available for conversion to fuels or energy per year. They estimated that about 194 million dry tons of biomass per year could be sustainably removed from agricultural lands today. This number includes 15 million dry tons of grains to biofuels. They do not include energy crops in the currently available number.

Table 4.3 summarizes the committee's estimate of the main components that make up the non-feed and food crop biomass available in the United States in the near term (today) at less than about $68 (2007 dollars) per ton, delivered to the processing plant. Also included in the table are other estimates for comparison. Crop residues are readily estimated (~60 percent recovery) and are available annually. Energy crops are estimated at 3 dry tons per acre per year on ~30 million acres of CRP land (which could be available in the near term). Forest residue is estimated at less than half of that projected to be available from logging residues and fire suppression (Perlack et al., 2005). Wood waste estimates are from Walsh et al. (2000) and Perlack et al. (2005). More than 95 percent of mill wastes are used internally already. Biomass costs, including transportation distance and cost, remain major uncertainties.

With future technology, Perlack et al. (2005) projected that today's technically available biomass amount could be *increased* to more than a billion dry tons per year within 35 to 40 years through a combination of technology changes (e.g., higher crop yields and improved residue collection technology), full adoption of no-till cultivation, and changes in land use to accommodate large-scale production of perennial crops. Forest residues were projected to increase by more than 225 million dry tons per year from today's use level, and by 88 million dry tons per year from today's estimated forest biomass availability to 368 million dry tons per year, through a combination of recovering wood wastes produced through traditional and improved logging operations, forest thinning for fire suppression, and increased productivity and

TABLE 4.3 Estimated Primary Solid Biomass Components Available in the United States in the Near Term and 2030 for Less Than About $65 per Ton (2007 dollars)

	Biomass Amount (million tons per year)				
Source	NRC Estimate[a]	Walsh et al. (2000)	Milbrandt (2005)	Perlack et al. (2005)	NRC 2030
Crop residues	160	50	173	179	315
Forest residues	55	44	62	136	55
Mill wastes[a]	5	90	88	106	
Urban wood waste	30	37	34	37	30
Energy crops	85	188	99	—	100
Total	335	509	456	458	490

[a]NRC estimate includes only that fraction that is estimated as not already being used.

use of forest growth beyond lumber requirements. Agricultural biomass availability increased even more than forest residue availability. Under their moderate-yield growth case, crop residue yields were assumed to increase up to 50 percent for corn and 100 percent for soybeans, and the efficiency of recovery increases from 60 to 75 percent over the next 40 years. Thus, recoverable corn stover today is estimated at 75 million dry tons per year at 50 percent recovery. Total stover biomass is therefore 150 million dry tons per year. This increases by 50 percent to 225 million dry tons of total stover produced, and when 75 percent of this is recovered, corn stover available for biofuel production is 170 million dry tons per year. Estimates for small grains and soybeans increase from 6 million dry tons per year (many current residue levels are too low to recover economically) to 26 dry tons per year. Recoverable wheat straw was estimated to increase from 11 million dry tons per year to 35 million dry tons. Grain quantities for biofuel production increase about 70 percent. Perlack et al. (2005) project even more aggressive growth numbers in their high-yield version of this case. For example, corn stover increases to 256 million dry tons (versus 172 million) per year, and most other agricultural biomass sources increase similarly. Without perennial crops, the projected amount of sustainable agricultural-based biomass production ranges from 425 million to almost 600 million dry tons per year (including 56 to 97 million tons each year of grains to biofuels) by 2050, assuming the high-technology changes summarized above (Perlack et al., 2005). Perennial energy crop production is assumed on 60 million acres and woody energy crop production is assumed on 5 million acres, with significant yield increases for each included. These perennial energy crops contribute between 156 million and 377 million dry tons of biomass per year. Their high-technology case involves significant land use changes raising questions of storage or loss of carbon from the soil, which could affect the CO_2 impacts of biofuels (see below). Figure 4.4 shows the projected maximum technically available biomass in the United States in 2050 under the high-technology scenario

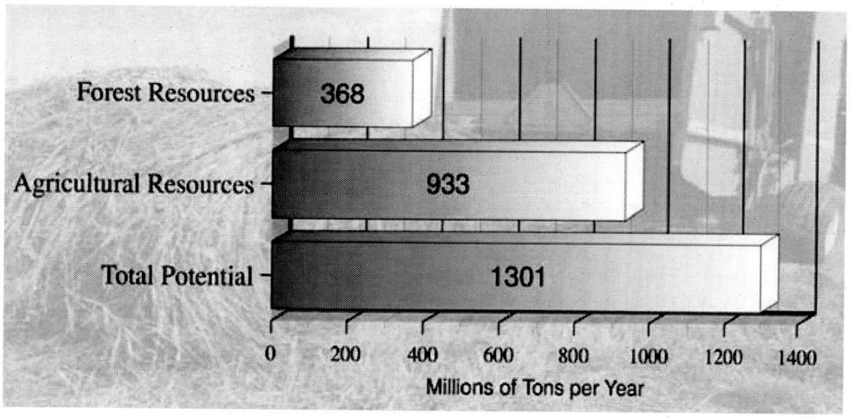

FIGURE 4.4 Projected sustainable biomass technically available in the United States by 2050, with aggressive energy crops. SOURCE: Perlack et al. (2005).

with aggressive production of energy crops (Perlack et al., 2005). Although these numbers are for the technical amount available from the listed sources, the cost can be assumed to be as high as $100 per dry tonne (2007 dollars) for some portion of this supply. Most studies, including that of Perlack et al., estimate technically available biomass as a function of cost per tonne, typically to about or exceeding $100 per dry tonne, and different biomass types have different cost profiles. At these prices, particularly by 2050, both algae and alternative crops for biodiesel, which can be grown on land unsuitable for food crops, may add to available sources. However, at $100 per dry ton for feedstocks, biofuels are unlikely to be competitive with other options as long as carbon emissions are cost-free.

Environmental issues are also associated with massive, intensive production of biomass, particularly via more traditional agriculture. These are not well quantified today but could involve increased depletion of soil nutrients and micronutrients and potential reduction of soil carbon, resulting in increased CO_2 emissions, significantly increased soil erosion, reduction in habitat and biodiversity, and increased stress on available water resources. Today, agriculture accounts for about 13 percent of greenhouse gas emissions, and land use changes account for about 19 percent (WRI, 2008). Land use changes frequently result in a loss of soil carbon; agricultural practices on cropland can either increase or decrease the amount of soil carbon (Tilman et al., 2007; Fargione et al., 2008). Thus, significant soil carbon loss can result in biofuels being a significant CO_2 emissions source. The above issues will become increasingly important as the amount of biomass directed to energy and biofuel production increases.

After reviewing significant research on the major issues with biomass supply for biofuels including water availability, cost, and avoiding significant land use changes, the committee estimates that about 500 million dry tonnes per year of biomass could be sustainably produced in 20 years at reasonable cost. This involves using crop residues at twice current levels (see Table 4.3); using switchgrass, mixed prairie grasses, and other energy crops at what is considered a conservative yield on CRP lands (about 3 tons per acre per year on 30 million CRP acres); and using forest or wood wastes at a moderate level at a level projected to be available today (Table 4.3).

More specifically on the basis for the increases in Table 4.3, corn yields, both grain and stover, have been increasing at nearly 1.2 percent a year recently (Arkebauer et al., 2004), and if this continues at 1 percent per year, with increased recovery efficiency, stover recovery will double. Less aggressive increases for the remainder of the crop residues, increased recovery efficiency, a <10 percent increase in acreage planted, and other agricultural residues produce the projected biomass amount. The energy crop estimate is considered conservative in light of recent data showing an average of 5.5 to 8 dry tons of biomass pre acre per year for the majority of 23 locations across the United States (McLaughlin and Kszos, 2005; Perlack et al., 2005). The committee judges that the combination of changes to achieve these production levels appears achievable.

Beyond 2030, if the sustainable amount of biomass obtainable from forest lands, estimated at 200 million dry tons per year, is added to the estimated sustainable biomass available from agricultural lands (~500 million dry tons per year), the total annually available biomass for biofuels by 2050 is estimated to be about 700 million dry tons per year. This is less than the 1.3 billion dry tons estimated by Perlack et al. (2005) because of other energy uses for biomass such as power generation and expected cost and water limitations. As noted earlier, the quantities of biomass available as a function of price remain unclear.

Production: Grain- and Sugar-based Ethanol

Most of the ethanol produced in the United States today is produced from corn via either dry milling or wet milling production processes. In dry milling, corn is ground into a fine powder, and the resulting meal is mixed with water and the enzyme alpha-amylase, which breaks down the starch into its individual sugar (glucose) components. The mixture is then heated to kill undesirable bacteria, the mash is cooled, and a second enzyme, glucoamylase, is added to convert the glucose into dextrose. Yeast is added and ferments the dextrose to ethanol and CO_2. The resulting mixture has a relatively low ethanol concentration in water because the yeast is inhibited when the ethanol content reaches a certain level (~13 percent ethanol). Ethanol is separated from water by distillation, which consumes a large amounts fossil fuel, typically natural gas, to distill off the water. The unconverted solid material consists of the corn's protein and nutrients, which upon drying makes good animal feed. Wet milling is similar to dry milling except that the corn kernels are soaked in weak sulfurous acid for about 48 hours before grinding. The ground mixture is then separated into solid and aqueous components. The starch-containing liquor is hydrolyzed enzymatically, then fermented, and finally distilled as in dry milling. Wet milling is somewhat more expensive because of the additional processing but it is amenable to larger-scale plants. Typical fermentation times are about 48 hours; fermentation times beyond 72 hours typically experience an increased failure rate because of contamination with bacteria, acetogens, or other contaminants.

Ethanol production in Brazil starts with sugar cane. The sugar is pressed out of the cane and fermented. The crushed cane left over from the sugar removal (called bagasse) provides heat for the process and the distillation, effectively eliminating the need for fossil fuels in the manufacturing process and the resultant net CO_2 emissions. Sugar cane can be grown much less expensively in Brazil and other tropical countries than in the United States, and Brazilian sugar-based ethanol has the lowest production cost of any biofuel, with costs running from \$0.80 to \$1.40 per gallon on a gasoline-equivalent basis (Paustain et al., 2006).

One bushel of corn can yield about 2.8 gallons of ethanol. To produce 3 billion gallons of ethanol per year requires about 1.1 billion bushels of corn, which in turn requires about 8 million acres. The United States currently produces about 11 billion bushels of corn annually on about 85 million acres. Thus, production of 3 billion gallons of ethanol per year consumes about 10 percent of our current corn crop. Increasing ethanol production to 6 billion, 9 billion, or 12 billion gallons per year would consume 20, 30, or 40 percent of our current corn crop, respectively. In view of the fact that we currently consume about 150 billion gallons of gasoline per year in our light-duty vehicle fleet, it is clear that corn ethanol cannot significantly impact oil imports for gasoline production, especially since ethanol contains less energy per unit volume, and much of that energy is in fact from fossil fuel. It seems unlikely that we could devote more than about 25 percent of our corn crop to ethanol because of the substantial cost increases in corn prices that would result and the impact of these higher prices across our economy and economies worldwide. Current experience is that the price of corn doubled when ~20 percent of the crop was directed to ethanol production. The U.S. Department of Agriculture (USDA) states that with projected increases in yield and corn acreage the United States should be able to produce about 11 billion gallons of ethanol per year while meeting food, feed, and export demands first (USDA-OCE, 2005). Thus, even with a 60 percent increase in corn production (increased yield and increased acreage), to move beyond 10 billion to 12 billion gallons of ethanol per year would require substantial ethanol imports, which could come from Brazil currently and in coming years potentially from a wide range of countries in Africa, Asia, and South America. Ethanol imports are currently discouraged by an import duty of \$0.54 per gallon on ethanol for fuel use. This would most likely have to change if imports are to make substantial contributions to our biofuel supply. Furthermore, today there are only limited volumes of ethanol available globally, and many countries are bidding for it. Planning on large volumes of imports raises the issue of large land use changes in other countries, which carries with it other environmental issues, including the potential for increased CO_2 emissions.

In 2002, with corn at \$2.36 per bushel, natural gas at \$3.00 per million British thermal units (Btu), and electricity at \$0.04/kW$_e$h, the cost of corn ethanol production was estimated at about \$1.20 per gallon of ethanol, including return on capital (Whims, 2002). Unleaded gasoline on the Nymex in that time frame was about \$0.80 per gallon. In 2007, due in large part to demand for ethanol, corn prices rose above \$4.00 per bushel. Natural gas prices have also risen and are currently over \$8.50 per million Btu. At \$4.00 per bushel of corn and \$8.50 per million Btu of natural gas, the cost of producing ethanol is about \$1.70 per gallon of ethanol. If the total transportation cost to the blending terminal is \$0.25 per gallon, the cost of ethanol ready to blend is \$1.95 per gallon. If the \$0.51 per gallon excise tax credit is applied, the apparent cost of ethanol at the blending terminal is \$1.44 per gallon. Since ethanol contains 80,000 Btu per gallon, versus about 119,000 Btu per gallon for unleaded gasoline, the cost of ethanol on a gasoline energy-equivalent basis is about \$2.90 per gallon. If the excise tax credit is applied, the apparent cost of ethanol per gallon at the blending terminal on a gasoline energy-equivalent basis is about \$2.15. In April 2007, unleaded gasoline on the Nymex was about \$2.00 per gallon (crude price was about \$65 per barrel). As a consequence, due in part to the rise in corn prices attendant on higher demand, higher oil prices have not made ethanol competitive economically. Thus in 2007, the Nymex selling price of ethanol was about \$2.05 per gallon. The higher cost can be justified on the basis of the high blending octane value of ethanol, but there are volume limits on this. In other words, once the required octane rating of the blended gasoline is reached, it is not justified economically to blend more of the higher-cost material into the pool because it only increases the cost of the fuel.

Cellulosic Ethanol

To address the ethanol volume issue as well as greenhouse gas emissions, biomass cellulose is a more attractive route because of its larger volume potential and because most of the feedstock is not in the food or feed chain. The stems, stocks, and leaves of plants and the trunks of trees (wood) are all composed of hemicellulose, cellulose, and lignin. Hemicellulose is a polymer of several sugars. Cellulose is a polymer of the six-carbon sugar glucose. These sugars can be fermented if the individual sugar units can be broken out of the hemicellulose and cellulose polymer chains. This is the challenge that must be overcome for cellulose- and hemicellulose-based biofuels to reach commercial viability via the biochemical route. Lignin is a highly cross-linked polymer that does not contain any sugar components but can be used as a renewable fuel for energy needed in processes used to convert the biomass to a liquid fuel.

Relevant biomass that could be a feedstock for cellulosic ethanol production can be essentially anything that grows. However, every biomass type is different and would be expected to offer different challenges for conversion. For illustrative purposes, consider the biomass associated with corn, with the exception of the grain. This biomass includes corn stocks, leaves, husks, and cobs and is referred to as corn stover. The corn stover is available only about one month per year and must be collected, bailed, and delivered to a central processing site. It must undergo washing and grinding for size reduction before it is introduced into the process.

The cellulose-based process is more complex than that for ethanol production from grain, particularly because of the need for acid hydrolysis (or another processing step) to break up the complex cellulose-hemicellulose-lignin structure before enzymes can free sugar molecules from the cellulose for fermentation, which is effectively the third major step. Each of these steps typically involves separate vessels. Further, the total residence time in the process is about twice as long as for grain ethanol production. These add capital and operating cost to the process. The increased hydrolysis and fermentation times can also lead to increased contamination and fermentation problems and to increased cost. Although the costs of the hydrolytic enzymes have been reduced tenfold or more over the last decade, this remains the key area in need of R&D attention. Cellulase is currently estimated to cost between \$0.10 and \$0.30 per gallon of ethanol produced (D'Aquino, 2007). A major development required for cellulosic ethanol to become competitive with grain ethanol and potentially with oil-based fuels is the development of new, more robust organisms that can hydrolyze cellulose and ferment the freed sugars in the same vessel at acceptable rates. This would reduce capital and operating costs. At this point there is no clear evidence that such robust organisms have been developed. Organisms that could operate at higher concentrations of ethanol would also reduce cost. Furthermore, different improved, more robust hydrolysis and fermentation cultures also require development.

Costs

Although the plant equipment can currently be specified and designed, a relatively large uncertainty in the production cost and the technical readiness will remain until the technology has been demonstrated on a commercial scale. Furthermore, the increased complexity, greater number of vessels, and longer residence times associated with cellulosic ethanol, compared to grain ethanol, production means that the cost of ethanol via the cellulosic route would significantly exceed that for the grain-based route. Using the above parameters, the estimated ethanol production cost is around \$2.50 per gallon. Paustain et al. (2006) estimate the cost of a gallon of ethanol to be \$1.95 ± 0.65 for biomass costing \$35 per dry ton. While DOE targets for "mature" cellulosic biomass conversion technology project parity with fossil fuels for transportation, achieving such parity represents a large stretch for the technology, which is yet to be demonstrated commercially.

Production Volumes

The potential ethanol production from corn stover can be estimated from the expected yield per ton of biomass of a cellulosic ethanol plant. Using the Perlack et al. (2005) estimate of 75 million dry tons of corn stover currently available per year and 60 gallons of ethanol per dry tonne leads to 4.5 billion gallons of ethanol from corn stover per year. For a typical plant size of 2,000 dry tons per day, 110 plants would be required to produce this amount of cellulosic ethanol from corn stover. This is a similar number of plants to that in place for grain ethanol.

From the total estimated currently available crop residue of 160 million dry tons per year (see Table 4.3), including corn stover, and a yield of 60 gallons of ethanol per dry ton, it would be possible to produce about 10 billion gallons of ethanol per year. If process improvements increase the conversion efficiency to 90 gallons of ethanol per dry ton, 14 billion gallons of ethanol could be produced per year. If by 2030, crop residues total 315 million dry tons per year and sufficient forest residues and perennial crops are added to reach a total sustainable annual biomass production of 500 million dry tons, the ethanol production potential from cellulosic materials could reach about 40 billion gallons per year at an average yield of 80 gallons per dry ton. Producing 35 billion gallons of biofuels by 2017, the current administration and legislative target, is a large stretch that is probably not achievable (see Case 3 in Chapter 6) without major biofuel imports. Further, without the development of new organisms that can simplify the cellulose conversion process, reducing capital costs and residence time, cellulosic ethanol will remain significantly more expensive to produce than grain ethanol, requiring significant subsidies to be economically attractive.

Biobutanol

Biobutanol is another potential entrant into the automotive biofuel market. Biobutanol is a four-carbon alcohol (versus the two-carbon alcohol ethanol). Several technologies to produce it are in the R&D phase. The one receiving the most attention is the acetone-butanol-ethanol (ABE) process. As currently envisioned, this process involves the bioconversion of sugars or starches from sugar beets, sugar cane, corn, wheat, or cassava into biobutanol using a genetically engineered microorganism, *Clostridium beijernickii* BA101. The midterm goal is to start with cellulose, but this awaits the success of the economic conversion of cellulose and hemicelluloses into sugars. Biobutanol has many attractive features as a fuel. Its energy content is close to that of gasoline; it has a low vapor pressure; it is not sensitive to water; it is less hazardous to handle and less flammable than gasoline; and it has a slightly higher octane than gasoline. Thus, it can go directly into the existing distribution system and substitute directly for gasoline. Its main drawback to date is its cost. To attack the cost and initiate market entry, DuPont and BP have joined forces to retrofit an existing bioethanol plant to produce biobutanol using DuPont-modified biotechnology (Chase, 2006). An improved next-generation bioengineered organism is projected to be available within a few years. The promise of biobutanol is to start with cellulosic biomass as a feedstock. The cellulose approach is being studied but is far from commercial.

Biodiesel

Biodiesel is a renewable fuel produced from vegetable or animal oils and fats. It is made by the transesterification of these feedstocks, typically with methanol. The reaction is catalyzed by a base such as sodium hydroxide (NaOH). The main products are a methyl ester, which is referred to as biodiesel, and glycerol, which has a variety of other uses (see Figure 4.5). If ethanol is used instead of methanol, the product is an ethyl ester; methanol has been preferred because it is cheaper than ethanol. Biodiesel has about 93 percent of the energy content per gallon of oil-based diesel fuel and has a cetane number between 50 and 60, with 55 being typical.

Cost of Biodiesel

Driven by the feedstock cost, which is about 80 percent of a plant's operating cost, the cost of biodiesel has been significantly higher than the cost of conventional diesel. Operating costs (excluding feedstock costs) are \$0.40 to \$0.45 per gallon. The variable cost (operating + feedstock, excluding capital) of producing biodiesel has been in the range of \$2.50-\$2.80 per gallon for the last decade, with soybean oil around \$2.00 per gallon (Hofman, 2003). This cost is for the use of methanol for esterification; if bio-based ethanol were used, the cost would be higher, but net CO_2 emissions would be reduced. The largest cost component in this is for the feedstock, which for soybean oil has been around \$2.00 per gallon of biodiesel. Increasing grain prices have increased the cost of soybean oil also. Yellow grease (used cooking oil) is about \$1.00 per gallon but is limited in volume. Outlays to cover capital and other fixed costs and returns to investors are likely to be more than \$0.20 per gallon, bringing the total cost of biodiesel to around \$3.00 per gallon for traditionally priced soybean oil (Hofman, 2003). Wholesale diesel fuel during this period was about \$1.80 per gallon, which means that biodiesel was not competitive. However, biodiesel from yellow grease was cost competitive. In 2007, soybean oil rose to \$3.00 per gallon, and the cost of biodiesel was about \$3.50 per gallon, whereas the cost of conventional diesel was about \$2.10 per gallon. In spring 2008, soybean oil rose to about \$4.20 per gallon (Brasher, 2008). The current viability of the industry depends on government incentives, which are about \$1.00 per gallon (Brasher, 2008), and programs.

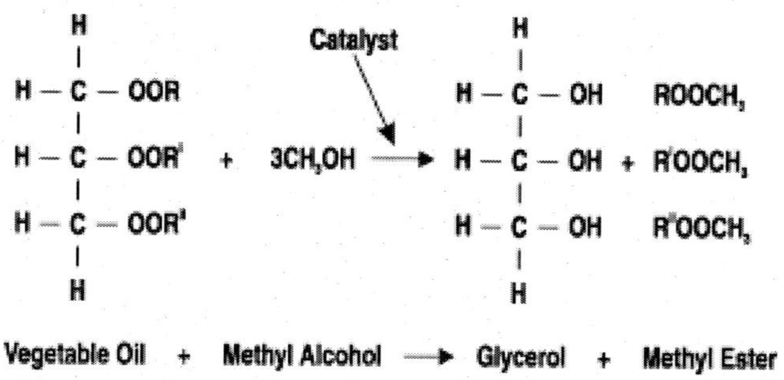

FIGURE 4.5 Transesterification of vegetable oils.

Production of Biodiesel

U.S. production grew from about 500,000 gallons per year (33 barrels per day [bbl/d]) in 1999, to 2 million gallons per year in 2000, to 250 million gallons per year in 2006, and 450 million gallons per year in 2007 (Brasher, 2008). This compares with 690 million gallons per year in Germany and about 200 million in France in 2006. For most of the period up to 2006 in the United States, the utilization rate of installed capacity was less than 25 percent. In 2006, 250 million gallons were produced from an installed capacity of about 580 million gallons per year (about 45 percent capacity utilization [Hofman, 2003]). In Europe, the industry has grown rapidly, with capacity expanding in step with demand. The United States is also in the process of building massive additional capacity, with 57 plants under construction or in the planning stages as of January 2008 (Brasher, 2008). Further, South America and Asia are building excess capacity in expectation of large export markets (Weirauch, 2006). Countries in these regions are studying new plants such as jatropha, which can be grown in areas unsuitable to traditional crops.

To produce 1 billion gallons of biodiesel from soybean oil would require about 690 million bushels of soybeans, 22 percent of our recent annual soybean crop of about 3.1 billion bushels per year. As with corn ethanol, an expansion in soybean oil use for biodiesel beyond perhaps 35 percent of our soybean production is likely to cause significant economic ripples through food and agricultural markets. If we are to go beyond the level of 1.5 billion gallons of biodiesel per year, we will most likely need to depend substantially on imports. Yellow grease availability will limit biodiesel production from this material to about 100 million gallons per year (Radich, 2004). In the case of biodiesel, current technology is relatively straightforward and well proven. Because of the simplicity of the process, there is not much to be gained in terms of economies of scale or process optimization. Forward projections are fairly linear and depend on the availability of animal and vegetable oils. The major issue is cost. The other option involves producing diesel fuel via biomass gasification, pyrolysis, or Fischer-Tropsch synthesis. This is considered in "Gasification Routes to Biofuels" below in this chapter.

Biodiesel has the advantages that it has good requisite diesel fuel properties (excluding low-temperature properties) and can be blended into the fuel supply, where desired, to utilize the existing infrastructure. Because production is simpler and less energy intensive than for corn ethanol, the use of biodiesel, on an energy-equivalent basis, can reduce CO_2 emissions per gallon by about 10 to 50 percent, depending on the plant oil source and the amount of fossil fuels used in growing, harvesting, and producing the biodiesel compared to a gallon of conventional diesel on a life-cycle basis (UK Department for Transport, 2008). For soybean oil the CO_2 reduction is about 30 percent. Biodiesel can reduce oil consumption by 90 to 95 percent relative to oil-based diesel usage (Radich, 2004).

Future Biosynthetic Biofuels

Over the last 25 years significant research efforts have focused on algae for biofuel production. Algae can be grown in both salt- and freshwater environments, in shallow ponds, tubes, or raceways utilizing waste nutrients. One area of research is the development of algae that have high lipid productivity (Briggs, 2004; Pacheco, 2006). The oil would be extracted from the collected algae. Progress was made early in the period, but efforts have slowed greatly. Recent reevaluation suggests that current costs are well over \$4.00 per gallon, and much more progress is needed if this technology is to have an impact in the foreseeable future (Pacheco, 2006). Alternatively, algae can be grown as a source of cellulose for biofuels.

With the rapid growth of synthetic biology and the increasing ability to engineer organisms to produce specific chemical or fuel products, the area of genetic engineering of biomass has witnessed significant renewed interest (Savage, 2007). Microbes that produce and excrete specific hydrocarbons minimize the energy-consuming separation costs that are the key issue with ethanol formation. Furthermore, properly designed hydrocarbon products in either the diesel range or the gasoline range could fit into the existing infrastructure without requiring new infrastructure as needed for ethanol at larger scales. Although no specific processes can be considered for commercial evaluation at this point, the rate of change and potential specificity that can be expected could produce rapid changes in the not-too-distant future. It is difficult to project future developments; thus, this technology is not considered further here, but it deserves careful tracking in the future.

Gasification and Pyrolysis Routes to Biofuels

Biomass gasification involves the conversion of biomass to a mixture of carbon monoxide, hydrogen, carbon dioxide, methane, and other organics including bio-oils and tars, ash, and small char particles. The concentration of these gases and other materials depends on the process design and operating conditions of the gasifier. Gasification has the advantage that it can convert essentially any biomass to clean synthesis gas from which a broad range of products can be derived. Biomass gasification exhibits many similarities to coal gasification including a significant number of gasifier types and different approaches to gasification technology. Fuels produced via gasification of biomass should have low net CO_2 emissions, and if gasification is combined with capture and sequestration of CO_2 emissions, such processes should have a negative CO_2 emission footprint. They should also result in an equivalent reduction in oil use.

Gasification is carried out under a variety of pressure and temperature conditions. When relatively low pressures are used, the process is basically rapid pyrolysis. Under low-pressure rapid pyrolysis conditions, primary products are a mix of hydrogen, CO, light hydrocarbons, bio-oil, and char. The heating is *usually indirect and operation is at lower pressure*, avoiding the need for an expensive air separation unit. The mix of primary products can be separated into several fractions for upgrading. Bio-oil, which is a complex mixture of organic compounds, can theoretically be refined further under process conditions that resemble oil refining. If this can be done commercially, it would provide an avenue to biomass-based liquid transportation fuels that would substitute directly for their oil-based counterparts. However, the requisite refining technologies need to be developed and demonstrated. In addition to deriving products from the bio-oil, the full set of gaseous and liquid pyrolysis products can be gasified, the gas stream can be compressed, the CO shifted to the appropriate CO-to-hydrogen molar ratio, and the CO_2 removed. The resulting synthesis gas can be converted to transportation fuel (e.g., diesel fuel via Fischer-Tropsch synthesis). Another option is to produce methanol from CO and hydrogen. Methanol can then be converted to dimethyl ether (DME) or to gasoline using methanol-to-gasoline (MTG) technology, which has been commercially demonstrated in a 14,000 bbl/d plant in New Zealand that operated successfully from 1985 to 1996. MTG appears to be the most likely route because MTG gasoline blends directly into the gasoline pool and fully uses the existing infrastructure. DME would require a separate or greatly expanded infrastructure.

Gasification using *direct firing with oxygen at higher pressures* produces a relatively pure stream of CO and hydrogen, with some CO_2 and other gases. This syngas mixture is easily shifted to the desired CO-to-hydrogen molar ratio and, after CO_2 removal, can be converted to diesel fuel by the Fischer-Tropsch synthesis or to methanol and then to gasoline by MTG. This approach would have the highest capital cost but, because of the simplicity of fuel production and a high yield, could be competitive.

Several U.S. and European groups are developing advanced biomass gasification technologies, and there are about 10 different biomass gasifiers with a capacity greater than 100 tonnes per day operating in the United States and worldwide. There are more than 90 installations (most are small) and 60 manufacturers of gasification technologies (BTG, 2004). For example, at the McNeil Generating Station in Vermont, a low-pressure wood gasifier, which started operation in August 2000, is converting 200 tons of wood chips per day into fuel gas for electricity generation. Many of the gasification technologies have technical or operational challenges associated with them, but most of these problems are probably resolvable or manageable with commercial experience. The most persistent problem area appears to be biomass feed processing and handling, particularly if a gasifier operates at high pressure and/or must contend with different biomass feeds. DOE has funded five other advanced biomass R&D projects (DOE-EIA, 2007).

Although several of the available gasification technologies have been demonstrated commercially, the technology is not yet commercially proven for biomass gasification and the production of biofuels. The technology is still on a relatively steep learning curve, and the integration of gasification, gas cleanup, and biofuel synthesis or bio-oil refining is yet to be commercially established. The same can be said for biomass gasification for hydrogen production. A major difference with biomass gasification is that it will involve smaller units than coal gasification and will not see the economies of scale of coal gasification. This will increase the cost per unit product unless major process simplification and capital cost reduction can be achieved. The primary approach to this is to eliminate the air separation unit, which is typically required with most high-severity gasification technologies, and its cost. This leads to gasification by indirect heating, which could more appropriately be considered biomass pyrolysis, as discussed above and typically produces bio oil, tar, and gases.

Several economic studies of biomass gasification have been reported, at times combined with demonstration projects. A Finnish company has estimated that its biomass gasification technology, which is operating at the 150 kW_{th} feed scale, can produce diesel fuel for \$2.50 to \$3.10 per gallon (VTT, 2006). Jensen (2004) from Shell estimated the cost of Fischer-Tropsch diesel from biomass gasification to be about \$1.80 per gallon. Using a consistent basis for comparison, rapeseed methyl ester biodiesel was estimated at about \$4.50 per gallon. These two sets of estimates are for large-scale plants. Choren Industries is in the early operating phases of a gasification process coupled with Fischer-Tropsch technology from Shell to produce high-quality diesel from a 50-50 mixture of wood chips and wood waste in Germany (Choren Group, 2008). This facility will demonstrate the gasification, syngas cleanup, and synthesis technologies integrated together.

The World Bank (2007) made a series of estimates for biomass-based integrated gasification combined cycle (IGCC) plants that were about one-fifth the size of typical coal-based IGCC plants. The size was limited by the maximum distance at which it was economic to harvest and transport biomass fuel. The estimated total plant cost was \$2,150/kW_e (1997 dollars) for a 100-MW_e wood gasification-combined cycle power plant. For a 32 MW_e wood gasification-IGCC demonstration plant in Brazil the total plant cost was about \$2,900/kW_e (1997 dollars). This compares to an estimated total plant cost of \$1,300/kW_e (1997 dollars) for a 500 MW_e coal-based IGCC. These results demonstrate that biomass gasification suffers from diseconomies of scale because of feedstock limitations. The higher capital cost associated with biomass gasification, combined with the higher cost of biomass feedstock supply, will make it difficult to compete with coal if carbon capture and storage (CSS) is not required.

Impact of Biofuels on Oil Imports and Greenhouse Gas Emissions

In part due to competition with the food and feed needs of the United States and the world, corn ethanol has severe limitations with respect to its potential volume and impact on oil imports and on reductions of CO_2 emissions from the transportation sector. Corn production in the field is energy intensive, using fossil fuels for fertilizer production, cultivation, and harvesting. The production of ethanol also requires large amounts of energy, which typically comes from natural gas, with the result that net energy and net CO_2 reductions are limited. Life-cycle estimates of the net energy of corn ethanol range from a loss to a more than 30 percent energy gain over the energy in the fossil fuels used in its production, depending on the assumptions made in the studies and on the system boundaries (i.e., what is included in and not included in the analysis). Figure 4.6 shows the range of results found in the literature. The specifics of the studies leading to these results are referenced and discussed by Farrell et al. (2006). A reasonable conclusion for grain ethanol is 18 to 25 percent energy gain over fossil inputs (Wang, 2005) and a similarly sized net CO_2 emissions reduction relative to gasoline.

Figure 4.7 shows the primary energy inputs of fossil and other energy sources in megajoules of primary energy and the estimated greenhouse gas emissions in kilograms CO_2 equivalent per megajoule of fuel for gasoline production and per megajoule of product energy for ethanol production, also from the work of Farrell et al. (2006). This shows that the use of dry-mill ethanol reduces CO_2 emissions by about 25 percent over the use of gasoline on an energy-equivalent basis.

Total fossil fuel use is also reduced by 20 to 25 percent (for dry-milled ethanol), but since most of the fuel used in growing, harvesting, and producing ethanol is natural gas and coal, the reduction in oil use that ethanol can produce is much greater (see Figure 4.7). Thus, dry mill ethanol use could reduce the petroleum requirement by about 0.95 gallon per energy-equivalent gallon of ethanol used.

Net CO_2 reductions improve when biomass, such as bagasse or lignin, supplies the energy needed for the conversion and separation. Thus, ethanol from cellulosic feedstocks can produce a roughly 88 percent reduction in net CO_2 emissions on a fuel energy-equivalent basis (see Figure 4.5) (Wang, 2005; Farrell et al., 2006). Estimated emissions from today's conventional vehicles are shown in Table 4.4. Because the cellulosic ethanol process is self-sufficient in terms of energy, the oil reduction achieved when cellulosic ethanol is used to displace a gallon of gasoline is about 0.93 gallon of oil on an energy-equivalent basis.

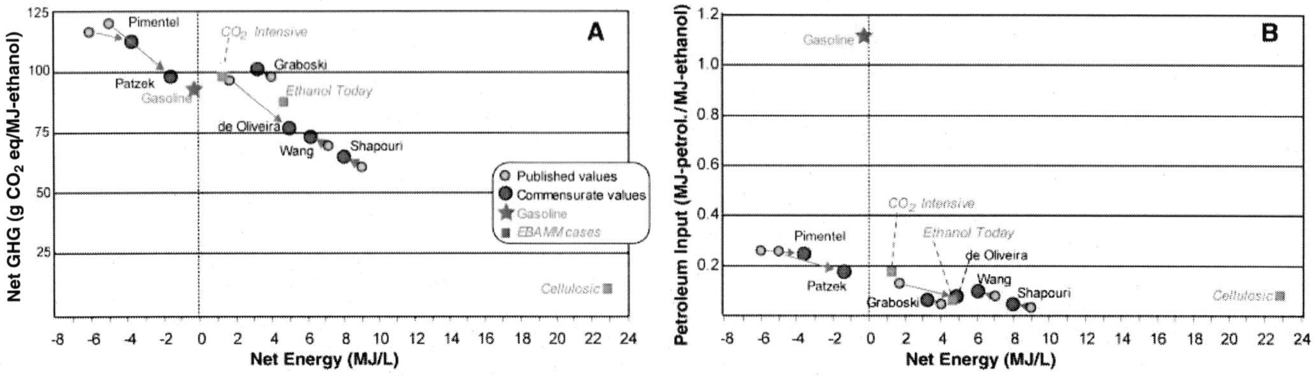

FIGURE 4.6 Published estimates of range of impacts on net greenhouse gas (GHG) emissions (A) and oil inputs (B) for grain-based ethanol. For additional discussion and information, see http:www.sciencemag.org/cgi/content/full/sci;312/5781/1748b. SOURCE: Farrell et al. (2006). Reprinted with permission of the American Association for the Advancement of Science.

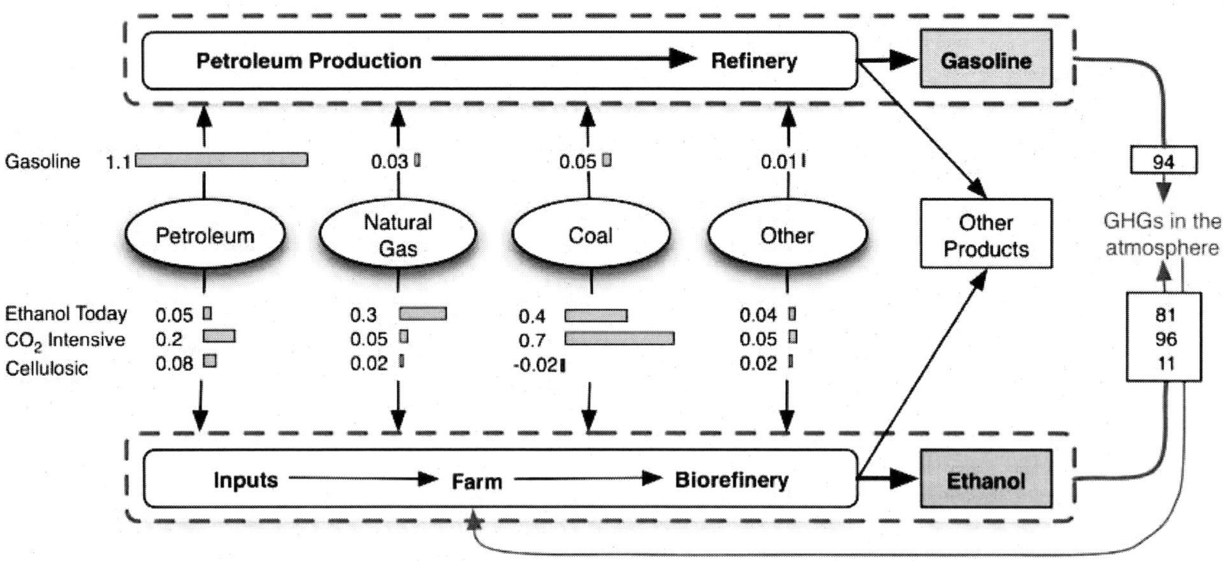

FIGURE 4.7 Primary energy inputs and net greenhouse gas (GHG) emissions for gasoline and ethanol (primary energy inputs are in megajoules per megajoule of product energy; GHG emissions are in kilograms of CO_2 equivalent per megajoule of fuel). SOURCE: Farrell et al. (2006). Reprinted with permission of the American Association for the Advancement of Science.

TABLE 4.4 CO_2 Emissions from Today's Conventional Light-duty Gasoline and Diesel Engines in a Typical Family Sedan and from Fuels from Less Conventional Sources

Vehicle Technology or Fuel Source	Well-to-Wheels CO_2 Emissions (g CO_2/km)
Conventional gasoline 2005	165
Conventional diesel 2005	143
Ethanol from corn	131
Ethanol from cellulose	15

SOURCE: Heywood (2007).

Production Potential of Biofuels

Production of ethanol from grain is fully commercial. Figure 4.8 shows the corn-ethanol production capacity growth from 1990 to 2007. U.S. production capacity grew from 4.3 billion gallons annually at year-end 2004, to 5.9 billion gallons annually by year-end 2006, and about 7 billion gallons annually by year-end 2007. Considering only current plant construction under way, ethanol capacity will be at least 8 billion gallons per year by year-end 2008 and could be as much as 10 billion gallons per year if all proposed projects are completed. This is a doubling of capacity in 4 years and

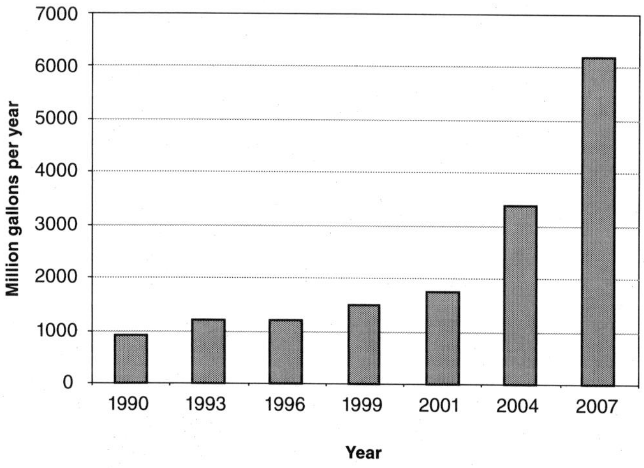

FIGURE 4.8 Growth in production of corn-based ethanol in the United States. SOURCE: Courtesy of Renewable Fuels Association.

TABLE 4.5 Key Assumptions and Parameters Used in Biomass-to-Biofuels Scenarios

	Probable Case	Upper-bound Case
Biomass potentially available in the near-term (dry tons/yr)[a]	335[b]	335[b]
Biomass available in 2050 (dry tons/yr)	500	700
Average ethanol yield on biomass, 2008 to 2030 (gallons/dry ton)	60	60
Average ethanol yield on biomass, 2030 to 2050 (gallons/dry ton)	90	90

[a]Biomass use starts with crop residues (corn stover) and then adds other sources including energy crops as they become available, driven by cost.

[b]See Table 4.3 for primary biomass components that make up this number.

is well ahead of mandated biofuel levels. However, the 10 billion to 12 billion gallons per year probably represents the limit with respect to corn availability, assuming that corn yields and acreage increase modestly. Production of ethanol from grain is not likely to see significant additional process cost reduction. Further, costs are not subject to economies of scale because current plant size is about at the limit and water use limitations will probably increase costs in future plants. Based on the previous analysis in this chapter, a forward projection of grain ethanol production is as follows:

- 8 billion gallons in 2008,
- 10 billion gallons in 2010, and
- 12 billion gallons in 2015.

Production of ethanol from cellulose is yet to be demonstrated at a commercial scale, and significant questions remain about the economic and commercial viability of the technology. Within the next 3 to 5 years, four to five commercial demonstration plants are expected. These will provide valuable information on cost, technology robustness, and particularly, commercial viability at the scale required to achieve large-scale cellulosic ethanol production. This information should be available by 2010. The committee expects the commercial and economic viability of cellulosic ethanol to remain a key issue for some time.

Cellulosic ethanol or other alternatives (e.g., biomass gasification, liquid-fuel synthesis) will be required to meet biofuel demand beyond levels achievable with grain ethanol. If commercially successful, cellulosic ethanol production could approach the growth experienced for grain ethanol. Cellulosic ethanol plants are similar to grain ethanol plants although somewhat more complex, and because of the dispersed nature of biomass, they will be limited in size to about two to four times that of ethanol plants. For the rest of this discussion, the committee assumes that cellulosic ethanol is commercially demonstrated by 2010 and capacity begins to grow.[5] Two cases are considered; both assume that cellulosic ethanol is economically competitive or there are sufficient fuel subsidies to make it competitive with oil-based fuels so that production capacity is built with private capital. One case is considered to be a measured response to the need to replace oil-derived liquid transportation fuels and is called the probable case. The other case involves a more aggressive application of the technology to generate liquid transportation fuels and is called the upper-bound case. Table 4.5 summarizes the key assumptions and parameters used for the two cases.

The *probable case* assumes a more measured pace of application of the technology in view of all the issues including process cost, water availability, biomass cost, other competitive uses of biomass and the ability to build plants and to increase biomass availability. For this case, the capacity build followed the grain ethanol capacity build experience, in which over a several-year period about 1 billion gallons of capacity was added per year. For this case (see Table 4.5 for details), the key assumptions are that the technology is commercially ready and there are 335 million dry tons of biomass available in the near term, increasing to 500 million dry tons available per year for conversion to biofuels by 2050. The *upper-bound case* also starts with 335 million dry tons of biomass available per year and assumes that by 2050,

[5]The DOE roadmap on cellulosic ethanol is "to accelerate cellulosic ethanol research, helping to make biofuels practical and cost-competitive by 2012." The three to five demonstration plants that DOE is funding should have achieved that goal by the end of 2010 or have identified the key remaining issues. If an economic business case can be made by that time, there is capital ready to build forward.

700 million dry tons of biomass are available per year and converted to biofuels (ethanol). It assumes that many of the constraints are resolved, that sufficient additional biomass is available, and that once sufficient experience is gained, plant capacity build exceeds the grain ethanol experience. It is consistent with the aggressiveness of the Hydrogen Success scenario (Case 1) discussed in Chapters 3 and 6 and, thus, is the most logical case for comparison with the hydrogen cases. It also shows the maximum impact that biofuels can have on oil import reduction and on greenhouse gas emission reductions from light-duty vehicle transportation.

Biomass for biofuel production starts with agricultural crop residues, which are readily available today and then double over time. Grown-for-purpose energy crops and forest residues contribute later in the period, driven by price, location, and local availability. Because of the issues of gathering and transporting biomass, conversion plants will be limited in size, potentially two to four times as large as corn-based ethanol facilities and will not be able to gain further economies of scale. Ethanol yield is assumed to improve from 60 gallons per dry ton early in the build to 90 gallons per dry ton about mid period and beyond. Water will become a major issue as the production grows. For forest residues and energy crops grown on less desirable land, the areas where plants will be built will likely be dryer, and water is expected frequently to become a major issue. This and other factors, including the cost of biomass per dry ton, will limit full utilization of the technically available biomass. The upper-bound case removes or limits many of these restrictions

The infrastructure build for rapid growth in cellulosic ethanol production should be similar to that experienced for corn ethanol. The primary difference is that cellulosic ethanol plants will cost two to three times as much as corn ethanol plants due to greater complexity and increased material and construction costs. The infrastructure for delivery of the product to blend plants would be expected to remain largely the same as for grain ethanol, but it would have to be expanded to meet the greater production. Since the plants are relatively small and widely scattered, truck and rail would probably pick up most of the growth. Dedicated pipelines connecting areas of substantial production with large blend plants would be expected to develop in specific instances. Costs can be expected to decrease but will remain a significant issue. The estimated build-out and projected annual cellulosic ethanol production for each case are given below:

Probable Case	*Upper-bound Case*
1 billion gallons in 2011	1 billion gallons in 2011
5 billion gallons in 2015	6 billion gallons in 2015
12 billion gallons in 2020	16 billion gallons in 2020
18 billion gallons in 2025	28 billion gallons in 2025
32 billion gallons in 2035	44 billion gallons in 2035
45 billion gallons in 2050	63 billion gallons in 2050

Biobutanol should be commercially demonstrated in 2 to 3 years with the BP-DuPont plant noted above, but the main issue will be cost. It will probably require about a decade to define the extent to which costs can be reduced, for next-generation microbe technology to be developed and applied, and for technology developed to convert cellulosic biomass to biobutanol. This is when commercial viability will be clearer. Biobutanol is included in the scenario by assuming that it is produced at 10 percent of the cellulosic ethanol production level and is offset by 5 years. Thus, 0.1 billion gallons per year could be available in 2015, 0.5 billion gallons per year in 2020, and so forth.

Biodiesel production from plant and animal oils is fully commercial, and the technology is considered mature. The production cost is mainly in the plant oil cost, and the yield of oil per acre is low for crops suitable for the United States. This suggests that it will not be a major player in the U.S. light-duty vehicle fuel market. U.S. production costs are about $3.00 per gallon. The technology is simple and mature, and there has been no problem building plant capacity to date, with recent utilization of installed capacity at less than 50 percent.

Projected annual production volumes are as follows:

- 250 million gallons per year today,
- 600 million gallons in 2012, and
- 1.5 billion gallons in 2020 (maximum production).

The 1.5 billion gallons of production per year would consume about 30 percent of the soybean crop and probably cannot be increased. Corn and soybeans compete for the same land, limiting soybean production.

Because the manufacture of biodiesel is much less energy intensive, the estimated impact on CO_2 emissions is a 30 to 35 percent reduction per gallon of biodiesel used, and the impact on oil consumption is about a 90 to 95 percent reduction relative to use of a gallon of regular diesel fuel. CO_2 emissions include those associated with the methanol used in esterification. The energy contents of biodiesel and regular diesel are roughly the same.

Conclusions on Biofuels

CONCLUSION: Although use of corn- and oil-based biofuels can provide some benefits in reducing U.S. oil use and CO_2 emissions, cellulosic biofuels will be required for such benefits to be significant. Lower-cost biofuel production methods and conversion processes will have to be developed for large-scale commercialization, but the initial high costs of biofuels, together with other barriers, may limit their market potential, absent policy interventions or significant oil price increases or supply disruptions.

Biofuels offer the potential to reduce oil imports because they can replace a fraction of the liquid fuels needed for U.S. light-duty vehicle transportation. They can also reduce CO_2 emissions because they use carbon that was captured by plants in their last growth cycle, not carbon stored during previous millennia, and the repeated growth cycles recapture the CO_2 emitted during combustion of the fuel. Biofuels from different sources will have a different impact on oil imports and on net CO_2 emissions.

Grain ethanol has a 20 to 25 percent energy gain over the fossil fuel inputs used for its production and, on average, reduces CO_2 emissions by 18 to 25 percent over the use of gasoline on an energy-equivalent basis. Grain ethanol production is fully commercial but is constrained by grain availability because it competes with the use of grains for food and animal feed.

Much more biomass is available from non-grain sources. The technology for cellulosic ethanol has not yet been demonstrated for commercial production. It should be significantly better than grain ethanol with respect to CO_2 emission reductions because plant lignin and other plant residues can be used to supply the needed manufacturing process heat, reducing the use of fossil fuels. The key issues for cellulosic ethanol are commercial readiness, economics, and sustainability of biomass production including maintenance or improvement of soil productivity.

Biomass gasification is technically feasible, and many components have been commercially demonstrated. If the CO_2 produced in gasification were captured and sequestered, biomass gasification would have a negative CO_2 emissions balance. Economics will be the primary issue with biomass gasification, as it is with cellulosic ethanol, but it currently appears to be more competitive and closer to commercial reality. For a given amount of biomass, the thermochemical routes will produce roughly the same amount of biofuels on an energy-equivalent basis as the biochemical routes. A potential route to biofuels is through the use of algae, either as a source of cellulose or as a means to produce hydrocarbons. These routes have yet to be demonstrated above the pilot scale.

Biodiesel production from plant and animal oils is fully commercial, and the technology is considered mature. The production cost is mainly in the plant oil cost and remains uncompetitively high.

Therefore, biofuels offer significant potential to reduce CO_2 emissions from oil use by the U.S. light-duty vehicle fleet. The extent of these reductions is highly dependent on the biofuel source. Grain-based ethanol and biodiesel are severely limited by grain availability and cost. Non-grain biomass offers a large new source of biofuels, but cellulosic ethanol technology is not commercially ready and gasification routes to biofuels require commercial demonstration. Either or both of these routes could greatly expand biofuels production. However, their availability and commercial timing are such that it appears highly unlikely that the United States will be able to meet the biofuels target of 35 billion gallons per year by 2017.

OVERALL CONCLUSION

CONCLUSION: The committee's analysis indicates that at least two alternatives to HFCVs—advanced conventional vehicles and biofuels—have the potential to provide significant reductions in projected oil imports and CO_2 emissions. However, the rate of growth of benefits from each of these two measures slows after two or three decades, while the growth rate of projected benefits from fuel cell vehicles is still increasing. The deepest cuts in oil use and CO_2 emissions after about 2040 would come from hydrogen. See Chapter 6.

BIBLIOGRAPHY

Achenbach, J. 2007. So What's So Bad About Corn? As Iowa Enjoys a Bumper Crop, Farmers Hear It from Environmentalists, Ethanol Skeptics and Other Critics. *Washington Post*. November 23, p. A1.

An, F., J.M. DeCicco, and M.H. Ross. 2001. *Assessing the Fuel Economy Potential of Light-Duty Vehicles*. SAE International, Warrendale, Pa.

Arkebauer, T.J., A. Dobermann, K.G. Cassman, R.A. Drijber, J.L. Lindquist, J.E. Specht, D.T. Walters, and H.S. Yang. 2004. Changes in Nitrogen Use Efficiency and Soil Quality after Five Years of Managing for High Yield Corn and Soybean. In *Fluid Focus: The Third Decade. Proceedings of the 2003 Fluid Forum*, Vol. 21. Fluid Fertilizer Foundation, Manhattan, Kansas.

Brasher, P. 2008. Troubled Biodiesel Plants Find Market in Europe. *Des Moines Register*, March 13, p. 1D-2D.

Briggs, M. 2004. *Widescale Biodiesel Production from Algae*. Available at www.unh.edu/p2/biodiesel/article_alge.html.

BTG (Biomass Technology Group). 2004. *Biomass Gasification*. Available at http://www.btgworld.com/.

Burke, M., G. Briffet, J. Fuller, H.H. Heumann, and J.P. Newall. 2003. *Powertrain Efficiency Optimization of the Torotrak Infinitely Variable Transmission (IVT)*. SAE International, Warrendale, Pa.

Caldwell, L. 2007. US Ethanol Forecast Identifies Refiner, Marketer Opportunities. *Oil and Gas Journal* 105(11):18-24.

Chase, R. 2006. DuPont, BP Join to Make Butanol: They Say It Outperforms Ethanol as a Fuel Additive. *USA Today*, Business Section, June 23.

Choren Group. 2008. Carbo-V Process for Producing SunDiesel. Available at http://www.choren.com/en/biomass_to_energy/carbo-v_technology/.

D'Aquino, R. 2007. Cellulosic Ethanol—Tomorrow's Sustainable Energy Source. *Chemical Engineering Progress* 103(3):8-10.

DOE (Department of Energy). 2007. *Historical U.S. Hybrid Vehicle Sales*. FreedomCAR Vehicles Technologies Program. Available at http://www1.eere.energy.gov/vehiclesandfuels/facts/2007_fcvt_fotw462.html.

DOE-EIA (Energy Information Administration). 2006. *Emissions of Greenhouse Gases in the United States 2005*. Washington, D.C.

DOE-EIA. 2007. *Electric Power Annual 2006*. Available at http://www.eia.doe.gov/cneaf/electricity/epa/epaxlfile1_1.xls.

Duleep, K.G. 2007. The Hydrogen Transition and Competing Automotive Technology. In National Research Council Panel Meeting on Fuel Cell Vehicles and Hydrogen. Washington, D.C.

DuPont. 2007. *Fermentive Production of Four-Carbon Alcohols*. World Patent Application.

EIA (Energy Information Administration). 2007. *Biomass for Electricity Generation*. Washington, D.C.

EPA (Environmental Protection Agency). 2005. *Light-duty Automotive Technology and Fuel Economy Trends: 1975 through 2005*. Washington, D.C.

EPA. 2006. *Light-duty Automotive Technology and Fuel Economy Trends: 1975 Through 2005*. Washington, D.C.

Fargione, J., J. Hill, D. Tilman, S. Polasky, and P. Hawthorne. 2008. Land Clearing and the Biofuel Carbon Debt. Available at www.scienceexpress.org. DOI: 10.1126/science.1152747.

Farrell, A.E., R.J. Plevin, B.T. Turner, A.D. Jones, M. O'Hare, and D.M. Kammen. 2006. Ethanol Can Contribute to Energy and Environmental Goals. *Science* 311(5760):506-508.

Heywood, J.B. 2007. Strategies to Reduce Transportation Petroleum Consumption and Greenhouse Gas Emissions. In FreedomCar and Fuels Program Meeting.

Hofman, V. 2003. Biodiesel Fuel. In *Extension Service*. North Dakota State University, Fargo.

Hucho, W.H. 1998. *Aerodynamics of Road Vehicles*. Fourth Edition. SAE International, Warrendale, Pa.

Jensen, P. 2004. Innovative Biofuel Production Processes: Fischer-Tropsch Synthesis and Hydro Thermal Upgrading. In *EU-China Workshop on Liquid Biofuels*. Beijing.

Kasseris, E.P., and J.B. Heywood. 2007. *Comparative Analysis of Automotive Powertrain Choices for the Next 25 Years*. SAE International, Warrendale, Pa.

Kromer, M.A., and J.B. Heywood. 2007. *Electric Powertrains: Opportunities and Challenges in the U.S. Light-duty Vehicle Fleet*. Massachusetts Institute of Technology, Cambridge, Mass.

McLaughlin, S.B., and L.A. Kszos. 2005. Development of Switchgrass (*Panicum virgatum*) as a Bioenergy Feedstock in the United States. *Biomass and Bioenergy* 28(6):515-535.

Milbrandt, A. 2005. *A Geographic Perspective on the Current Biomass Resource Availability in the United States*. DOE, Washington, D.C.

Nishigaya, M., T. Tamura, H. Yasue, S. Kasuga, and M. Sugaya. 2001. *Development of Toyota's New "Super CVT."* SAE International, Warrendale, Pa.

NRC (National Research Council). 2004. *The Hydrogen Economy: Opportunities, Costs, Barriers, and R&D Needs*. Washington, D.C.: The National Academies Press.

NRC. 2006. *Tires and Passenger Vehicle Fuel Economy*. Washington, D.C.: The National Academies Press.

NRC. 2008. *Review of the Research Program of the FreedomCAR and Fuel Partnership: Second Report*. Washington, D.C.: The National Academies Press.

Pacheco, M. 2006. The Potential of Biofuels to Meet Commercial and Military Needs. Presentation to the committee.

Paustain, K., J. Antle, J. Sheehan, and E. Paul. 2006. *Agriculture's Role in Greenhouse Gas Mitigation*. Pew Center, Washington, D.C.

Perlack, R., et al. 2005. *Biomass as Feedstock for a Bioenergy and Bioproducts Industry: The Technical Feasibility of a Billion-Ton Annual Supply*. DOE-USDA, Washington, D.C.

Radich, A. 2004. *Biodiesel Performance, Costs, and Use*. Available at http://www.eia.doe.gov/oiaf/analysispaper/biodiesel/index.html.

RFA (Renewable Fuels Association). 2008. Industry Statistics. *2007 Monthly U.S. Fuel Ethanol Production/Demand*. Available at http://www.ethanolrfa.org/ (March 2008).

Savage, N. 2007. Better Biofuels. *Technology Review* 110(4):1.

Tilman, D., J. Hill, and C. Lehman. 2007. Carbon-Negative Biofuels from Low-Input High-Diversity Grassland Biomass. *Science* 314(5805):1598-1600.

UK Department for Transport. 2008. *Carbon and Sustainability Reporting Within the Renewable Transport Fuel Obligation*. Table 5, p. 39 and pp. 184-193. London.

USDA-OCE (U.S. Department of Agriculture-Office of Chief Economist). 2005. USDA Agricultural Baseline Projections to 2014. February.

VTT. 2006. *VTT Launches Gasification Test Equipment for Development of Second-Generation Biofuels*. Available at http://www.vtt.fi/uutiskirje/042006art05.jsp?lang=en.

Walsh, M., et al. 2000. *Biomass Feedstock Availability in the United States: 1999 State Level Analysis*. Oak Ridge National Laboratory, Oak Ridge, Tenn.

Wang, M. 2005. Updated Energy and Greenhouse Gas Emissions: Results for Fuel Ethanol. In *15th International Symposium on Alcohol Fuels*. San Diego, Calif.

Weirauch, W. 2006. Biodiesel: Uneven Supply, Demand. *Hydrocarbon Processing* 85(12):7.

Weirauch, W. 2007. Examining the Roll of Ethanol. *Hydrocarbon Processing* 86(4):9.

Weiss, M.A., et al. 2000. On the Road in 2020: A Life-cycle Analysis of New Automobile Technologies. Massachusetts Institute of Technology, Cambridge, Mass.

Whims, J. 2002. *Corn Based Ethanol Costs and Margins*. Kansas State University, Manhattan, Kansas, pp. 1-11.

World Bank. 2007. *Biomass Gasification Technology Forecast*. Available at http://www.worldbank.org/html/fpd/em/biomass/igcc_appendix.pdf.

WRI (World Resources Institute). 2008. World Greenhouse Gas Emissions Flow Chart. Available at http://www.wri.org/chart/world-greenhouse-gas-emissions-flow-chart (accessed March 18, 2008).

5

Role of the Stationary Electric Power Sector in a Hydrogen Fuel Cell Vehicle Scenario

The U.S. stationary electric power system is composed of generating facilities that convert primary energy sources into electricity, transmit that electric power at high voltage over distances ranging from a few miles to hundreds of miles, and distribute it, at reduced voltages, to an array of customers ranging from residences to large, industrial complexes. The most important primary energy sources today are coal, oil, natural gas, nuclear fuels, solar, wind, and hydroelectric power.

Unlike hydrogen, which can be produced and stored, electricity must be produced instantaneously to meet the demand for electric power, because there are very limited viable methods for large-scale electrical energy storage. This difference may provide a useful mechanism for the production of hydrogen. Hydrogen and electricity do share an important characteristic—namely, both energy carriers are derived from other primary energy resources, another fact that may prove to be synergistic. Figure 5.1 shows schematically the various ways in which the stationary power sector can interact with the transportation sector.

In 2005 the nation's electric power system, owned by hundreds of investor-owned, cooperative, and government utilities was composed of 978 gigawatts (GW) of generating capacity and produced 4,055 (TWh; terawatt-hours (billions of kilowatt-hours) of electricity) (EIA, 2007). Because of the typical daily load cycle of the generation, there is a meaningful fraction of generation capacity that is not used in the off-peak period. One could not fully utilize all of that unused

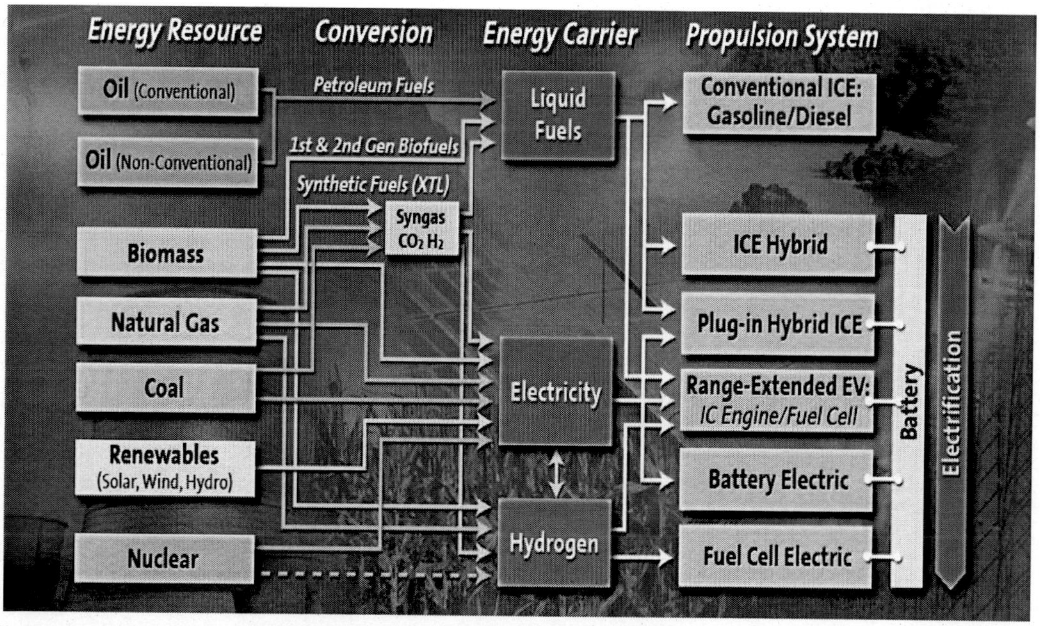

FIGURE 5.1 Stationary power and the transportation system. SOURCE: Beriesa (2007).

capacity for other purposes. However, as a measure of potential resource availability, off-peak capacity could in principle fuel more than a third of the light-duty fleet for a daily drive of 33 miles on average (Kintner-Meyer et al., 2006).

In considering the current and future electric power sector, there are three ways in which it could be a significant factor in a hydrogen fuel cell vehicle (HFCV) future: (1) hydrogen production, either through electrolysis or co-production with electricity, (2) synergies between fuel cells for transportation and stationary applications, and (3) use of electric power for battery-powered vehicles. On this basis, three groups of questions emerge:

1. To increase the hydrogen available for transportation by 2020 and/or 2035, what could be done in the stationary power sector to accelerate hydrogen production? What are the technological requirements? What incentives would help?
2. How best can we develop and accelerate the use of hydrogen in the stationary power sector by 2020 and/or 2035? Again, what are the technological requirements and incentives?
3. Is there a plausible alternative use of the stationary power sector's excess capacity and infrastructure that can result in a viable alternative to hydrogen use in transportation in 2020 and/or 2035?

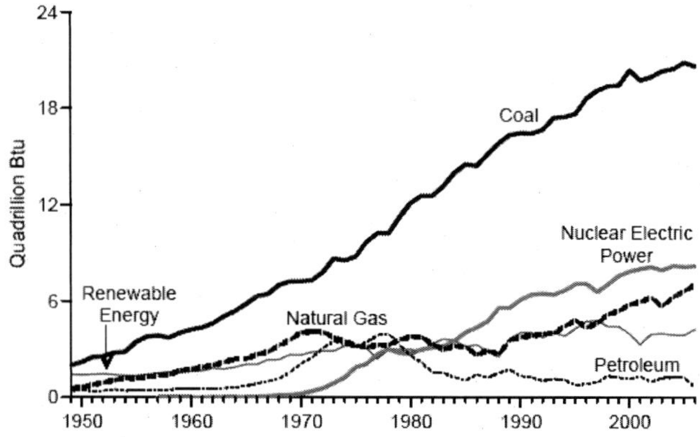

FIGURE 5.2 Energy source consumption for electricity generation. Renewable energy includes hydroelectric power. SOURCE: EIA (2007).

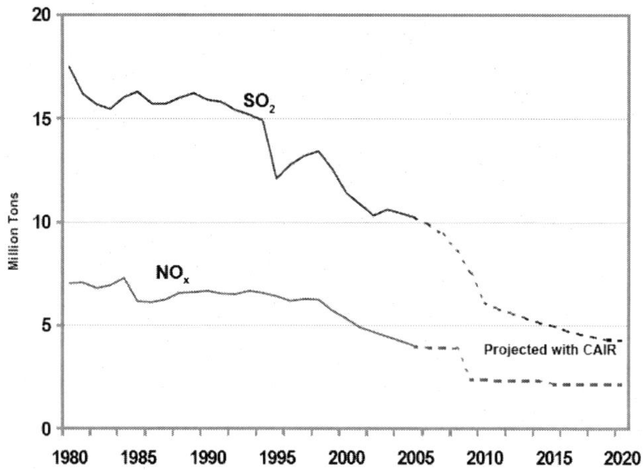

FIGURE 5.3 Nationwide NO_x and SO_2 emissions from the power sector. SOURCE: Srivastava et al. (2005).

TECHNOLOGICAL READINESS

It is useful first to examine the technological readiness of the systems mentioned earlier and the likelihood of their deployment in 2020 and 2035.

Hydrogen Production in the Power Sector

Producing hydrogen as a transportation fuel is somewhat similar to producing electricity for stationary use because both are energy carriers that require primary energy sources (mainly coal, nuclear, natural gas, and hydropower). About 40 percent of all energy used in the United States goes to producing electricity, which is the main form of energy in the residential and commercial sectors. Less than 3 percent of electricity is produced from oil as shown in Figure 5.2.

Furthermore, power plant emissions have declined significantly even though electricity demand continues to grow. This is shown for nitrogen oxides (NO_x) and sulfur dioxide (SO_2) in Figure 5.3. In the future, expected pressures for cleaner electricity production processes will continue the evolution toward low or zero emissions. In addition, the power industry is facing a significant challenge to reduce greenhouse gas (GHG) emissions in anticipation of a future carbon constraint. This has resulted in a move toward low or zero-carbon emitting technologies (i.e., renewable energy, nuclear energy, and fossil energy with carbon capture and storage). Tying hydrogen production to the industry's assets and processes could extend electric power benefits into the transportation sector, which is currently heavily oil dependent, with attendant pollutant and GHG emissions. Toward this end, an ad hoc group, the Hydrogen Utility Group, was formed in 2005 by nine power companies with the support of the Department of Energy/National Renewable Energy Laboratory, Electric Power Research Institute and National Hydrogen Association to explore the potential synergies between electricity and hydrogen production.

Hydrogen Production in the Power Sector: Electrolysis (Near Term)

As explained in Chapter 3, electrolysis (splitting water molecules to release hydrogen) is a proven, commercially available technology. Electrolysis, although more expensive than natural gas reformation, offers some unique features as an alternative method:

- Since the infrastructure to deliver electricity exists in every corner of the United States, hydrogen can be produced close to or at the point of distribution, thereby minimizing initial capital investments in hydrogen infrastructure.
- Unlike the reformation process where natural gas is the sole primary energy source, the energy sources for electricity production are diverse, including fossil fuel such as coal and natural gas; nuclear; and renewable energy, such as hydro, wind, solar and biomass.
- The cost of electricity has been and is expected to remain much more stable than natural gas prices due to fuel diversification in the power generation sector. (In specific regions, factors such as regulated-deregulated market and marginal cost of power generation for electrolytic hydrogen production could impact such comparisons.)
- Hydrogen produced by electrolysis emits no harmful substances at the point of production, which would be beneficial in environmentally sensitive urban areas. CO_2 emissions from power plants would be easier to capture and sequester than those from small natural gas reforming plants.
- Continuing progress to clean up power plant criteria and CO_2 emissions will ensure that hydrogen produced from electricity in the future will benefit from the increased cleanliness fossil fuel power generation.
- As a side benefit, the electrolysis process also produces oxygen that may have a market value.

As explained in Chapter 3, large electrolyzers using alkaline technology (producing more than 500 kg of hydrogen per day (kg/d) constitute a proven, commercially available technology. Smaller electrolyzers, using proton exchange membrane (PEM) technology, require more research, development, and demonstration (RD&D) to improve durability and efficiency and to reduce capital cost. For example, the capital cost per unit of production of a 10 to 100 kg/d PEM electrolyzer is four to seven times that of a 1,500 kg/d unit (EPRI, 2007). These smaller PEM units, if successfully developed, could provide an alternative or complementary approach to natural gas reformation for hydrogen production during the early commercialization stage.

From the power industry's perspective, the installed generation capacity of any utility is built to meet peak power demand. Thus, a portion of this capacity sits idle during off-peak periods, such as during the night when demand is reduced. This results in cycling of the power plants to respond to time-varying demand. While some plants are explicitly designed for peaking operations (e.g., combustion turbines), others are designed for base load operation (e.g., coal, nuclear). To the extent that base load plants are not used for base load operation, they must be cycled on a daily basis, and this carries a significant "wear-and-tear" cost penalty from cycling them up and down. If that capacity could be used to produce hydrogen during the off-peak period, via electrolysis, power plants could minimize such cycling and hence increase overall capacity factor and asset utilization. If 10 percent of all light-duty vehicles were fueled with hydrogen produced solely from electrolysis off-peak, the U.S. grid could realize an average of 8 percent increase in load factor. However, the electrolyzer plant would operate only about 50 percent of the time, and the increased capital charge per kilogram of hydrogen produced would to some extent offset the reduced power cost.

Today the electrolysis process is only moderately efficient, which makes it applicable only in certain niche markets when high-purity hydrogen is required. However, it would be beneficial to develop more efficient and cost-effective electrolysis technology. Capital cost reduction and improvements in electrolysis efficiency would be very useful, potentially making the economics of electrolysis more competitive with natural gas reformation.

Hydrogen Production in the Power Sector: Co-production (Long Term)

If HFCVs become widespread and hydrogen vehicle penetration increases in the longer term, large central hydrogen production facilities become more viable due to economies of scale. As described in Chapter 3, hydrogen production from coal with carbon capture and sequestration (CCS) could become one plausible way to meet the larger demand.

Integrated gasification-combined cycle (IGCC) power generation technology using coal is being developed, demonstrated, and commercialized. IGCC is more efficient in both power generation and emission control and, hence, could become the preferred alternative to pulverized coal power plants using a conventional combustion process. Furthermore, capturing and storing carbon dioxide (CO_2) produced from the coal in geologic formations 5,000 to 10,000 feet under the ground would reduce its impact on climate change, if long-term burial with minimal leakage can be achieved. Essentially complete storage is guaranteed in depleted or partially depleted oil and gas reservoirs, which have demonstrated long-term storage capability by the nature of their past storage capacity. Storage in aquifers, deep coal beds, and other formations is likely but yet to be fully demonstrated. Capturing CO_2 involves adding a water-gas-shift reactor to the IGCC process to convert the CO in the synthetic gas to CO_2 and hydrogen. CO_2 then would be separated from hydrogen, compressed, and sent underground via a pipeline for storage (sequestration), while the nearly pure hydrogen stream could enter a combustion turbine for power generation. The hydrogen produced in this process could be utilized for transportation fuel. This approach allows one to leverage a significantly higher upfront capital investment through the co-production of electricity and hydrogen, thereby making

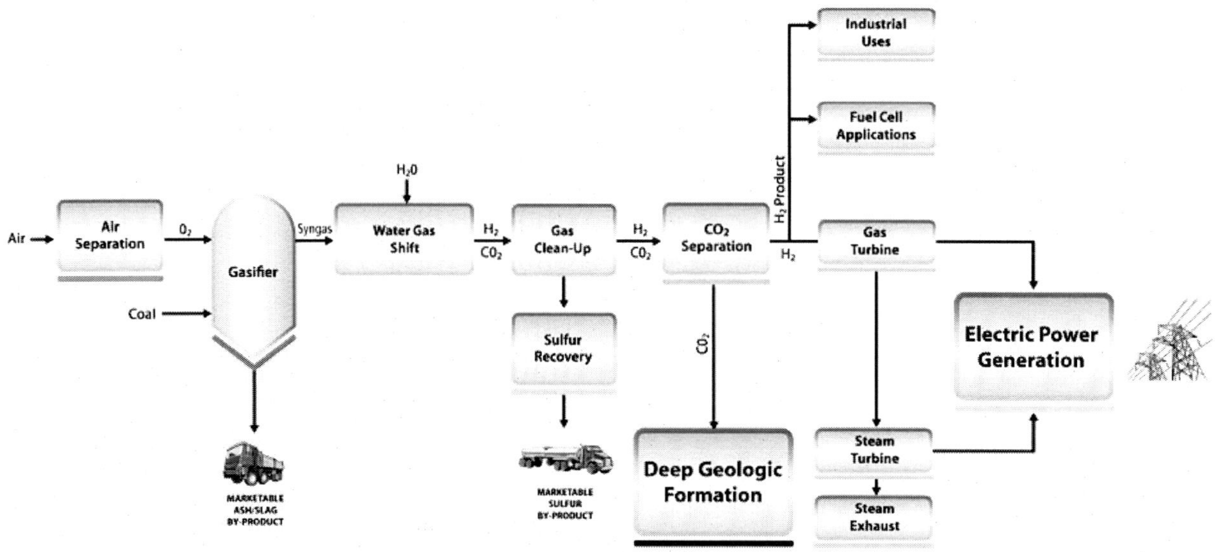

FIGURE 5.4 FutureGen concept for co-production of power and hydrogen.
SOURCE: FutureGen Alliance (http://www.futuregenalliance.org/).

both products more competitive because of shared costs for gasification and gas cleanup and operational flexibility (constant output of the syngas with swings between electric power and hydrogen production). A partnership between the Department of Energy and several industrial companies has been working to develop such a concept, FutureGen (Figure 5.4).[1] Success of a FutureGen-type power generation concept would establish a plausible pathway for both clean power generation and hydrogen co-production. The question that electric utilities (or independent power producers) will have to confront is to what extent utilities will be willing to invest in co-production facilities. While utilities may not want to enter the "new" hydrogen market, their reluctance could be lessened by policy actions by their state public utility commissions (PUCs) to provide incentives for them.

Nuclear power offers another alternative to support large-scale hydrogen production to meet high market demand in the long term. Nuclear power is receiving renewed interest since it produces neither harmful air pollutants nor greenhouse gases (although minor amounts of CO_2 are emitted in the fuel fabrication process). DOE has an active program to consider the co-production of hydrogen with advanced high-temperature gas-cooled nuclear reactors, but the time line for such development efforts is currently lagging the IGCC-CCS efforts and is unlikely to be ready to serve the hydrogen demand by 2025-2030 when centralized hydrogen production facilities are needed. As a result, it is not considered in this study. However, this does not diminish its potential beyond the 2025-2030 time frame to compete with or complement the IGCC-CCS technology.

Potential for Synergy from Large-scale Stationary Fuel Cells for Stationary Power

As discussed in Chapter 3, PEM is the technology being developed by all major vehicle manufacturers for primary power in their prototype fuel cell vehicles. In addition, PEM fuel cell systems are currently being developed for stationary applications, ranging from very small capacity backup power applications providing less than 1 kW to primary or stand-alone power applications of several hundred kilowatts. To the best of the committee's knowledge there are no PEM fuel cells systems currently in high-volume commercial production, although several companies have low rate commercial production and/or extensive field tests under way.

Many anticipated high-volume manufacturing target dates for stationary power have been missed, and potential consumers are now somewhat wary. Furthermore, many developers' initial product offerings target specialized, high-value, but relatively low-volume, market segments such as remote telecommunications or data center backup. Taken together, these considerations make it unlikely that stationary PEM fuel cell systems will precede vehicle fuel cell systems into

[1]DOE announced restructuring of the FutureGen project on January 30, 2008, citing cost escalation and technology advancement over the last 5 years since it was first announced in 2003. Under the restructuring plan, DOE intends to demonstrate the commercial viability of CCS technology at multiple commercial power plant projects that are either under way or in the planning stage. DOE will fund 100 percent of the incremental cost of the CCS portion of the projects if they are included in the plans, and it anticipates spending up to $1.3 billion (in as-spent dollars) between FY 2007 and FY 2020. These demonstrations should achieve the same technical specs as those of the original FutureGen plant with 90 percent CO_2 capture according the DOE's Request for Information document and, therefore, will have the same opportunity for hydrogen co-production.

the market in large-scale, mass market production by more than a few years, if at all.

Significant synergies between stationary and transportation PEM fuel cells during the time frame covered by this report might be expected in the area of ongoing product development, both in technology improvement and in manufacturing cost reduction. However, product requirements for PEM fuel cells in stationary and light-duty transportation applications are not the same. The DOE (2007) states on its website that the fuel cell stack cost target to be competitive with conventional technology for automotive is \$30/kW, whereas cost targets for stationary applications range from \$450 to \$700 per kilowatt for widespread commercial applications, with up to \$1,000/kW for some specific high-value, low-volume applications. The costs are higher because durability requirements for stationary applications are much higher; stationary fuel cells would operate for far more hours per year than those in automobiles. The DOE states that a 5,000-hour lifetime (approximately equivalent to 150,000 miles) will be required for fuel cell technology to be acceptable in automotive applications, while as much as a 40,000-hour lifetime may be required for widespread stationary power applications to be economically viable. Similarly, duty cycles, operational environment requirements, fuels, and other key performance parameters are likely to be quite different for automotive and stationary applications.

Stationary and transportation PEM fuel cells do share some common underlying development needs. The most important of these is reduction of installed costs. Estimates of anticipated costs for both stationary and transportation PEM fuel cell systems in volume production have been carried out by Battelle and the National Renewable Energy Laboratory (Kintner-Meier et al., 2006; NREL, 2005; Stone, 2005). These analyses show that in each system, most of the cost of a fuel cell system comes from the fuel cell stack itself. Thus, although fuel cell stack cost reduction or performance improvement specific to one application may not be directly transferable to the other, it is likely that general benefits would still accrue to both applications. For example, if transportation lifetimes were easily met with more stable long-life membranes developed for stationary applications, the "excess" lifetime capability might be used to benefit the transportation system in some other way, such as reducing materials costs and/or improving performance by using a thinner membrane.

Similarly, manufacturing process development for the PEM stack is likely to benefit both stationary and transportation applications. Some examples (not exhaustive) might include a better understanding of how to mold composite bipolar plates and/or better processes for forming and passivating metal bipolar plates, improved processes for production of membrane electrode assemblies (MEAs; e.g., membrane handling, application of catalyst, etc.), better methods for pre-testing MEAs or cells, and better methods for assembling the stacks themselves.

Beyond such basic stack activities, however, the picture is murkier. The product requirements, and thus the overall system designs, for the two applications are quite different. As a result, it is more difficult to identify significant synergies between stationary and transportation PEM systems either in detailed stack design or at the system level.

Long-term Potential for Synergy from Large-scale Stationary Fuel Cells for Stationary Power

Large (utility)-scale stationary power production currently accounts for about 38 percent (EIA, 2006) of the carbon emissions in the United States. As fuel prices rise and emissions standards are tightened, power equipment manufacturers spend increasingly large sums to achieve small efficiency gains (1 percent or less) and to reduce emissions. Further improvements of such gas-steam turbine combined-cycle systems (now a maximum of about 60 percent efficient) will be even more difficult to attain.

In the future, high-temperature fuel cells offer the possibility of further gains in efficiency and reductions of emissions, especially when operated in a hybrid mode with a turbine bottoming cycle to recover additional energy from the high-temperature exhaust and residual fuel stream exiting the fuel cell. Figure 5.5 shows a conceptual schematic for such a high-temperature hybrid fuel cell system. Theoretical fuel-to-electricity conversion for such systems approaches 70 percent.

Efficiency and emissions benefits in this system result primarily from the direct conversion of approximately two-thirds of the chemical energy in the fuel to electricity in the fuel cell stack, thus avoiding the efficiency limits gas and steam turbine and production of emissions associated with combustion processes. Further reduction of emissions also occurs due simply to the improved efficiency and resulting reduction in fuel consumption. It should be noted that such high-temperature fuel cell systems do not readily provide co-production of hydrogen for other purposes. The high-temperature fuel cell stack is able to use higher-order hydrocarbons (e.g., methane, natural gas) as fuel either directly or with limited external pre-reformation of the fuel. Thus, unlike the situation with low-temperature fuel cells (e.g., PEM), there is no place in the system where a ready source of pure hydrogen is available.

Several companies are pursuing such high-temperature fuel cell hybrid systems. However, significant technical challenges remain for these systems, and in the judgment of this committee it is unlikely that any significant reductions in oil imports or carbon dioxide emissions will result from widespread commercialization of these systems during the time frame studied in this report. Nevertheless, these systems do offer long-term potential for significant reductions in oil imports and CO_2 emissions and should form part of a portfolio of development activities that might potentially address these concerns.

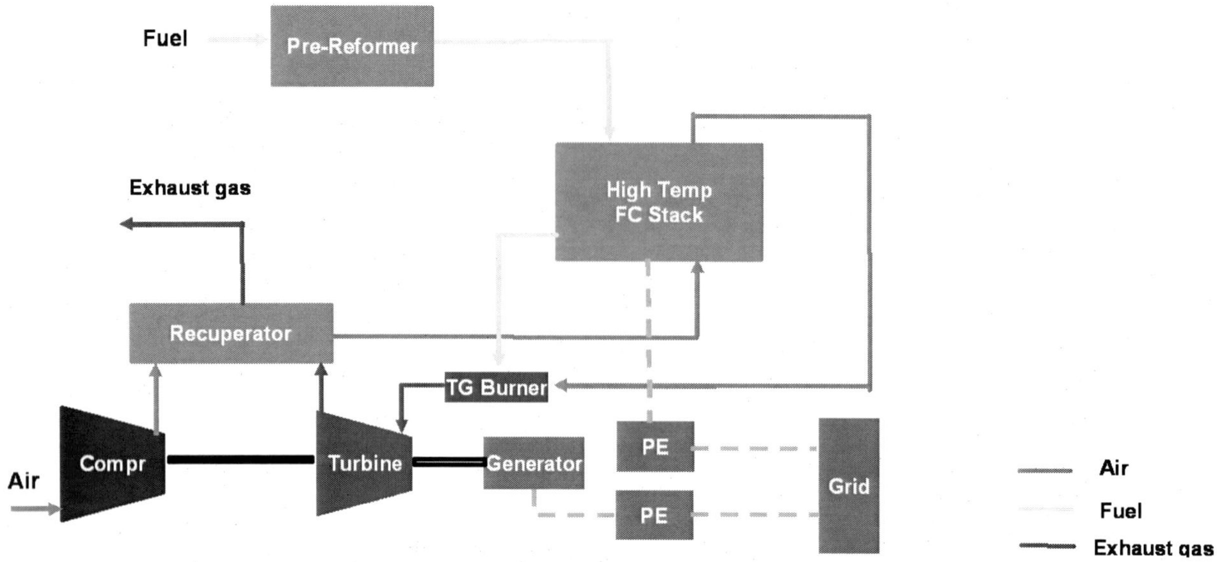

FIGURE 5.5 Schematic of high-temperature fuel cell hybrid system. SOURCE: General Electric Corporation, presentation to the committee.

Competitive Use of the Electric Power System for Electric Vehicles or Plug-in Hybrid Electric Vehicles

The stationary power sector could serve the transportation market in another way by providing electric power to charge batteries. These could then power electric vehicles (EVs) or plug-in hybrid vehicles (PHEVs). In this case, electric power, generally off peak at attractive prices, would be used to recharge batteries, normally at the vehicle's long-term "parked" location (home of a residential customer, garage for a fleet vehicle, etc.). PHEVs might have a range of about 20-40 miles. When the battery charge is depleted, a regular gasoline engine would start to operate the vehicle. PHEVs are described in Chapter 4.

Either type of electric vehicle would be much easier to implement than HFCVs, especially the PHEV. Little new infrastructure would be needed for the introduction of PHEVs, although new generating capacity and possibly transmission lines would be needed eventually. Infrastructure and logistics are much bigger problems for the introduction of HFCVs. EVs might require the construction of public charging stations to permit long-distance operation. The viability of both EVs and PHEVs depends on significant improvements in battery capability.

Two types of PHEVs are under consideration: the AER (all-electric range) and the "blended" PHEV. The AER has a large electric motor that provides all the traction power and a large battery to allow considerable all-electric vehicle operation. It also has a small engine that acts as a range extender when the battery is depleted. The blended PHEV has a larger engine and a smaller electric motor which operate in parallel to drive the wheels. The blended configuration is similar to current hybrids but with a larger battery to allow some operation on just electric power though less than the AER.

The use of off-peak power is particularly attractive. If off-peak power is available for \$0.07/kWh, it is equivalent to gasoline at \$0.77 per gallon after taking into account the differences in efficiency (Pratt et al., 2007).[2] As this example demonstrates, time-of-day pricing would have to be available to make off-peak power economical. This is an excellent example of how an electric utility, with the approval of its state PUC, could help make electric vehicles financially attractive. Similar incentives could be put in place for any plug-in or hydrogen-based concept.

The near-term focus should be on the blended version since the AER approach essentially has the same problems as full electric vehicles—namely, the need for a large, advanced battery and the need for rapid growth of charging stations. In a blended PHEV, the advanced batteries will be closer in size to those found in today's hybrid vehicles. With currently envisioned technologies, it would take an AER PHEV up to 6.5 hours to recharge (at 110 V) for a 40-mile battery range, whereas a blended PHEV would require only 2 hours (at 110 V) for a 5-mile battery range (Kawai, 2007).[3] It should be emphasized that the critical path developmental item for PHEVs is the advanced battery (e.g., lithium ion) that either the AER or the blended version will require.

The stationary power sector itself could supply the needs of about 40 percent of the current light-duty fleet for an average 30+ miles/day using off-peak power from current power

[2]It should be noted that gasoline prices include federal and state gasoline taxes, which are used to build and maintain roads.

[3]At 220 V, the charging times would be much shorter.

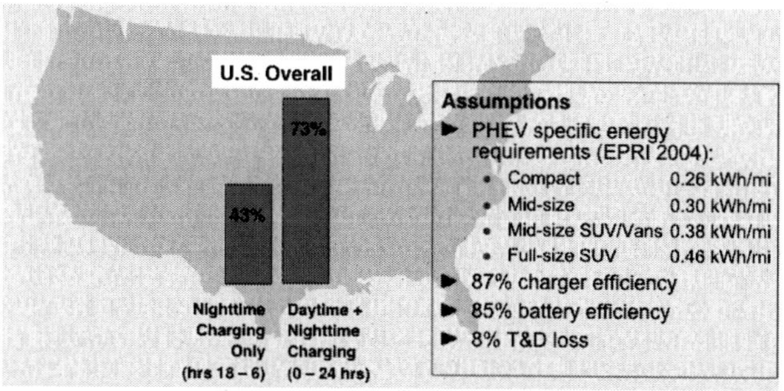

FIGURE 5.6 Fueling capacity for plug-in hybrid electric vehicles (PHEVs) in the U.S. power sector. SOURCE: Kintner-Meier et al. (2006).

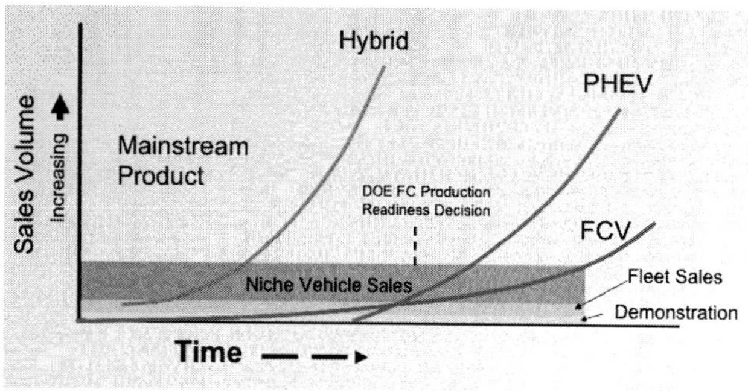

FIGURE 5.7 Advanced vehicle market penetration. SOURCE: Kawai (2007).

plants if PHEVs and EVs were available, equivalent to what it could do for HFCVs via electrolysis. The Pacific Northwest National Laboratory (Kintner-Meier et al., 2006) found that if all light-duty vehicles were plug-in hybrids, 70 percent or more of them could be charged if 24-hour per day charging were carried out utilizing otherwise reserve power generation capability. This relationship is shown in Figure 5.6.

In the scenarios presented in Chapter 6, the continued introduction of hybrid electric vehicles (some of which could be PHEVs) has been considered as an alternative to the introduction of hydrogen vehicles and the contrasts in infrastructure requirements are factored in. The most important factors are the time horizons under consideration (both 2020 and 2035) since the rollout of stationary hydrogen production (electrolysis and co-production) and stationary applications for fuel cells require time for technology development, permitting or regulatory approvals, and development of an infrastructure to support any such effort.

The EV and PHEV alternatives have a more limited, but also challenging, technological requirement, namely, the development of a high capacity battery to make this option viable. However, the permitting or regulatory approvals and infrastructure needs are much smaller, especially for the early, transitional period. This is shown in Figure 5.6 as well as in the qualitative estimate from Toyota (Figure 5.7).

INCENTIVES FOR THE ELECTRIC POWER SECTOR

The regulatory regime for electricity is well known and evolving. That regime can be extended to the hydrogen production (and plug-in hybrid) market without major changes except, possibly, for an incentive tariff.

One question, then, is what regulatory mechanisms might be put into place to provide utilities with incentives to produce hydrogen at either distributed or central locations. It is clear that additional revenues from hydrogen business and carbon emission credits would be reasonable incentives to make utilities consider branching out into hydrogen production, but would they be sufficient?

Let us start with the expected availability of the stationary electric power system to provide the power either for electrolysis (or PHEV charging) as well as the expected capacity for co-production of hydrogen and electricity by the new-generation options explained above.

In the stationary electric power sector, a variety of mechanisms have been used to encourage the introduction of new technologies or new approaches to doing business. These range from such federal approaches as production tax credits and carbon credits to state mechanisms such as rate structures and portfolio standards. The early introduction of such mechanisms would expedite action on the part of utilities to become players in this field sooner rather than later. This should be seriously considered, since utilities can be major players in the rollout of the systems described here.

The synergies for the use of stationary power for either hydrogen production and/or plug-in hybrids are quite significant. Improved asset utilization (increased capacity factors using electrolysis at distributed locations, i.e., substations and/or charging batteries) could (1) help increase generation capacity factors, (2) shave peak loads, (3) reduce wear and tear on cycling generation, and (4) provide hydrogen to transportation refueling stations or plug-in locations.

In the longer term, the co-production of electricity and hydrogen could leverage investments that would otherwise be required anyway, for electric power production. The co-production of hydrogen and electricity using IGCC technology with carbon capture and storage (as in the FutureGen concept) represents a potentially significant opportunity. Such an approach would leverage the capital investment since the fossil fuel, in this case coal, would be required to go through the gasification process, producing hydrogen that can be combusted to create electricity as well as producing hydrogen for transportation use. Through this process, carbon would be captured and sequestered.

CONCLUSIONS

CONCLUSION: With appropriate policies or market conditions in place, potential synergies between the transportation sector and the electric power sector could accelerate the potential for reduced oil use and decreased CO_2 emissions as benefits from the use of hydrogen in both sectors. In the near term, electrolysis of water at refueling sites using off-peak power, and in the longer term (after 2025), cogeneration of low-carbon hydrogen and electricity in gasification-based energy plants, are potential options that offer additional synergies. See Chapter 5.

More specifically, in response to the three framing questions posed at the beginning of this chapter, the committee reached the following conclusions:

1. In the near term (until 2020), existing electric power facilities (generation, transmission, substations, etc.) could produce hydrogen for transportation fuel purposes. In particular, small-scale electrolyzer plants, when successfully developed to meet more competitive cost and performance standards, at or near the points of distribution, could be important during the transition when the cost burdens of larger-scale reformation plants would be a potential barrier.
2. In the longer term (2035), the successful demonstration of one or more technologies could result in the widespread deployment of "co-production plants." One benefit from this approach would be the reduction in the use of natural gas that will increasingly have to be imported and is a source of greenhouse gases.
3. Incentives are likely to be necessary for full involvement of electric power companies. Mechanisms such as production tax credits, rate adjustments, carbon credits, and so forth, would be options for near-term action.
4. PEM fuel cell systems, whether for transportation or stationary systems, still require significant cost, reliability, and lifetime improvements to be truly competitive in the market. In many basic technology and product development issues, and basic manufacturing process development for the PEM stack, synergies between stationary electric power and transportation fuel cells might be realized.
5. The introduction of high-temperature fuel cells (solid oxide fuel cells or molten carbonate fuel cells) does not enhance the production of hydrogen since one advantage of these technologies is their ability to use a variety of feedstocks with an internal reformer.
6. Plug-in hybrid vehicles could help reduce reliance on imported oil (and natural gas). The reductions in overall CO_2 production will be a function of the reliance on fossil fuels for electricity production and the success of CCS technologies. The introduction of PHEVs depends on the timely introduction of advanced batteries.
7. The power industry could be of further assistance by providing a special electricity rate structure to support the early implementation of electrolysis. This is particularly important during the hydrogen market transition period. (See Chapter 3 for detailed discussions of the impact of the cost of electricity on the hydrogen production cost.) Utilities, working with their regulatory commissions could provide economic incentives to hydrogen producers to lessen the cost burden of the electrolysis process.

REFERENCES

Bereisa, J. 2007. Energy Diversity: The Time Is Now. Presentation to the committee, June 25.

DOE (Department of Energy). 2007. Fuel Cell Technology Challenges. Available at http://www1.eere.energy.gov/hydrogenandfuelcells/fuelcells/printable_versions/fc_challenges.html.

EIA (Energy Information Administration). 2006. Emissions of Greenhouse Gases. Available at http://tonto.eia.doe.gov/FTPROOT/environment/057306.pdf.

EIA. 2007. Annual Energy Review. Available at http://www.eia.doe.gov/emeu/aer/contents.html.

EPRI (Electric Power Research Institute). 2007. Electrolyzer Competitive Benchmarking: Pathways to Electrolysis-Powered Hydrogen Fueling Stations. Palo Alto, Calif.: 1014877.

Kawai, T. 2007. Sustainable Mobility and the Development of Advanced Technology Vehicles. Presentation to the committee, June 26.

Kintner-Meyer, M., K. Schneider, and R. Pratt. 2006. Impacts Assessment of Plug-in Hybrid Vehicles on Electric Utilities and Regional U.S. Power Grids. Part 1: Technical Analysis. Richland, Wash: Pacific Northwest National Laboratory. Available at http://www.pnl.gov/energy/eed/etd/pdfs/phev_feasibility_analysis_combined.pdf. Accessed September 2008.

NREL (National Renewable Energy Laboratory). 2005. Cost Analysis of PEM Fuel Cell Systems for Transportation. NREL Subcontract Report NREL/SR-560-39104. December.

Pratt, R., M. Kintner-Meyer, K. Schneider, M. Scott, D. Elliott, and M. Warwick. 2007. Potential Impacts of High Penetration of Plug-in Hybrid Vehicles on the U.S. Power Grid. Richland, Wash.: Pacific Northwest National Laboratory. Available at http://www1.eere.energy.gov/vehiclesandfuels/avta/pdfs/phev/pratt_phev_workshop.pdf. Accessed June 2008.

Srivastava, R., N. Hutson, and F. Princiotta. 2005. Reduction of Mercury from Coal-fired Electric Utility Boilers. Presentation at the DOE/NETL Mercury Control Technology R&D Program Review, Pittsburgh, Pa., July 12, 2005.

Stone, H.J. 2005. Economic Analysis of Stationary PEM Fuel Cell Systems. Battelle Memorial Institute, Project ID # FC 48.

6

Hydrogen and Alternative Technologies for Reduction of U.S. Oil Use and CO_2 Emissions

Estimating future transportation fuel use is difficult because of the complexities and uncertainties inherent in the analysis. Petroleum may continue to be the dominant fuel, or production may become constrained and prices rise much further. Hydrogen may replace petroleum as the main fuel, or it may not become significant at all. As discussed in Chapter 3, fuel cell vehicles and hydrogen have the potential to become competitive with conventional vehicles and fuels, but it is far from certain when that might be. Competitiveness depends in part on the cost of petroleum, which itself is highly uncertain, as witnessed by recent dramatic escalations in world oil prices. Nevertheless, as discussed in Chapter 2, there appear to be compelling reasons why the nation may have to reduce its use of petroleum, and hydrogen is among the leading candidates proposed to achieve dramatic reductions. Moving to a hydrogen-based transportation sector would be a revolutionary change that is unlikely to happen by itself. Mapping a route is essential to understanding how such a change might happen. Toward that end, this chapter formulates and analyzes several scenarios to map plausible futures for the use and impacts of hydrogen fuel cell vehicles (HFCVs) and other alternative vehicles and fuels. The scenarios and analyses necessarily depend on a host of assumptions. None of the scenarios should be viewed as projections of what the committee thinks is likely to happen. Rather, they are intended to describe different paths along which events may unfold and the consequences, especially for oil consumption and carbon dioxide emissions.

SCENARIOS AND ANALYSIS

Scenarios

The main object of the scenario analysis is to estimate the maximum practicable penetration rate of fuel cell vehicles, and then to estimate the resulting reductions of petroleum use and emissions of carbon dioxide (CO_2) in 2020 and beyond; the investments that would be needed during a transition period to bring hydrogen fuel cell vehicle technologies to cost competitiveness with gasoline vehicle technology; and the costs for a future hydrogen infrastructure. The committee developed three scenarios in order to investigate the range of possible outcomes. The hydrogen scenario analyses are based on the results presented in Chapter 3. In addition, as discussed in Chapter 4, hydrogen is not the only way to reduce petroleum use. Two scenarios focused on alternatives are analyzed, and a final scenario looks at combining all the approaches.

- Case 1 (Hydrogen Success) assumes that development programs are successful, as shown in Table 6.1, and that policies are implemented to ensure commercial deployment. Hydrogen fuel cell vehicles are introduced starting with a few thousand vehicles in 2012, growing to a fleet of almost 2 million by 2020, 60 million in 2035, and 220 million in 2050 (Figure 6.1). This rapid-growth case corresponds to a scenario recently developed by the U.S. Department of Energy (DOE) to 2025 (Gronich, 2007) and extended by the committee to 2050. By 2050, 80 percent of new vehicles sold are assumed to be HFCVs (Figure 6.2). This is consistent with other recent modeling studies (Greene et al., 2007).

—Case 1a (Hydrogen Accelerated) assumes that hydrogen and fuel cell vehicles are introduced at twice the rate of Case 1, although technical and cost goals are met at the same rate as Case 1. By 2020, 4 million hydrogen fuel cell vehicles are in the fleet. By 2050, 95 percent of new vehicles sold are assumed to be HFCVs This case is intended to investigate whether hydrogen could replace even more oil than in Case 1 if the nation faces a crisis situation, perhaps from declining worldwide petroleum production or rapidly worsening global climate change.

—Case 1b (Hydrogen Partial Success) assumes that developing programs fall short of the costs and performance of Case 1 (Hydrogen Success). Thus, the market introduction rate is slower than for Case 1, similar to DOE's "Scenario 1" (Gronich, 2007). By 2020, fewer than 1 million HFCVs

TABLE 6.1 Assumed Cost and Performance of Hydrogen Fuel Cell Vehicles and Gasoline Reference Vehicles[a]

	Case 1: Hydrogen Success	Case 1b: Hydrogen Partial Success	Reference Gasoline ICEV (AEO 2008)	Efficient Gasoline ICEV (Case 2)
FC drivetrain retail price (including fuel cell and hydrogen storage	Costs fall with learning and manufacturing scale to \$100/kW	Costs fall with learning and manufacturing scale to \$130/kW	\$54/kW	\$64/kW
HFCV retail price increment compared to gasoline reference vehicle	>\$100,000 (initially) → \$3,600 (learned out)	>\$100,000 (initially)→ \$6,100 (learned out)	—	\$1,000
FCV market Introduction	2012 (Case 1) 2010 (Case 1a)	2015	—	—
New car fuel economy[a]	51 mpgge (2015) → 85 mpgge (2050) = 2 × *efficient gasoline case*	45 mpgge (2015) → 73 mpgge (2050) = *1.75 × efficient gasoline case*	2005: 20.2 mpg 2015: 25.0 2050: 31.7	20.2 mpg 25.2 42.4

NOTE: Case 1a, Hydrogen Accelerated, is the same as Case 1, Hydrogen Success, for these values. Costs and fuel economy of HFCVs are based on a reference midsize vehicle with an 80 kW fuel cell. While small relative to most current engines, this would give equivalent performance, in part because of weight reductions in the body. This vehicle is assumed to represent the range of vehicles from small to large. Given all the other uncertainties in this analysis and the limited resources available to the committee, this assumption was both reasonable and unavoidable.

[a]On-road fuel economy is assumed to be 80 percent of the EPA average.

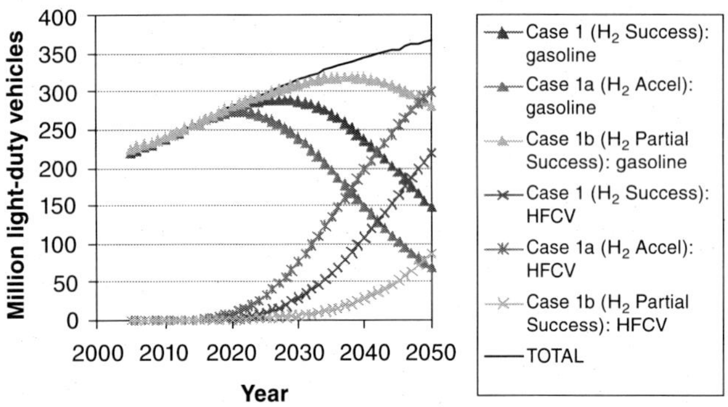

FIGURE 6.1 Hydrogen cases: Number of gasoline and hydrogen fuel cell vehicles in the fleet over time for three hydrogen cases.

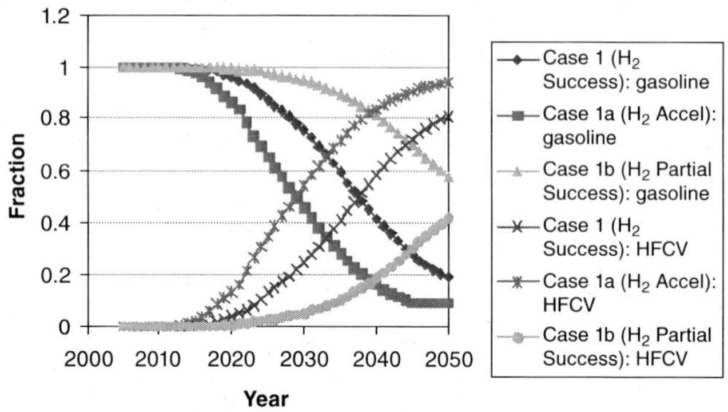

FIGURE 6.2 Hydrogen cases: Fraction of new gasoline and hydrogen vehicles sold each year.

are on the road, about the same rate of market penetration as hybrid electric vehicles have experienced.

- Case 2 (ICEV Efficiency) investigates the impact of improving conventional internal combustion engine ICEV fuel economy with currently feasible and expected technology. Fuel economy more than doubles by 2050, as shown in Table 6.1, in the committee's judgment the maximum practical rate with evolutionary vehicle technology. This analysis is based on the results in Chapter 4.
- Case 3 (Biofuels) examines the large-scale use of biofuels production from crop and cellulosic feedstocks. This level of production, equivalent to a maximum practical rate, is based on the results in Chapter 4.
- Case 4 (Portfolio), "all of the above," analyzes the impact if HFCVs, more efficient conventional vehicles, and biofuels are all pursued simultaneously.

TABLE 6.2 Hydrogen Supply Pathways Considered in This Analysis

Resource	Hydrogen Production Technology	Hydrogen Delivery Method to Station (for Central Plants)
Natural gas	Steam methane reformation (on-site)[a] Steam methane reforming (central plant)	N/A Liquid hydrogen truck Compressed gas truck Hydrogen gas pipeline
Coal	Coal gasification with carbon capture and sequestration (central plant)	Liquid hydrogen truck Compressed hydrogen gas truck Hydrogen gas pipeline
Biomass (agricultural, forest, and urban wastes)	Biomass gasification (central plant)	Liquid hydrogen truck Compressed hydrogen gas truck Hydrogen gas pipeline
	On-site reforming of ethanol	N/A
Electricity (from various electric generation resources)	Water electrolysis (on-site)	N/A

NOTE: N/A = not applicable.
[a]On-site refers to hydrogen production at the refueling station.

These scenarios are compared to a reference case, based on the Energy Information Agency (EIA) 2008 *Annual Energy Outlook* (EIA, 2008). The committee selected the AEO high-oil-price scenario for its reference case as being more representative of conditions under which HFCVs are promoted than the AEO reference case.[1] This scenario includes improvements of gasoline ICEV technology to meet CAFE (corporate average fuel economy) standards, although fuel economy continues to grow slowly after 2020, and some use of biofuels (blending up to 10 percent ethanol) but no introduction of hydrogen or advanced ICEV technology. Gasoline taxes continue as per AEO 2008.

The time frame for analysis is 2008 to 2050. The committee agreed that HFCVs were not likely to make a large impact on U.S. oil use and greenhouse gas emissions by 2020, because they are unlikely to enter the market before 2012-2018, and then it will take time to build up a large enough number of vehicles to impact oil use and carbon emissions significantly. The committee recognizes that uncertainties increase in such a long-term analysis, but it was necessary for examining the time frame during which hydrogen could have a large impact.

Technologies Considered

Hydrogen and fuel cell technologies are based on technologies currently in development, as discussed in Chapter 3 and recently reviewed in the National Research Council (NRC) FreedomCar Fuel Partnership report (NRC, 2008). Available hydrogen supply pathways are listed in Table 6.2. Hydrogen production or storage technologies that would require fundamental scientific breakthroughs (for example, hydrogen storage in carbon nanostructures or biological production of hydrogen by algae) are not considered.

Cost and performance data for current and midterm (2015-2030) hydrogen infrastructure technologies are discussed in Chapter 3. Efficiency improvements in ICEVs and biofuels are described in Chapter 4.

Modeling Tools for Scenario Analysis

The committee developed two EXCEL-based models for infrastructure and scenario analysis:

1. Hydrogen infrastructure model: designs and costs hydrogen infrastructure to meet a specified market penetration for HFCVs.
2. Simplified transition model: estimates investment to bring HFCV costs to competitive levels, investment costs for building hydrogen infrastructure, oil savings, and greenhouse gas emission reductions,[2] over time.

The models were developed at the University of California at Davis (UC Davis) and are described in detail in Appendix C. Given the time and resources available to the committee, the models were of necessity relatively simple, but they

[1]In this scenario, imported low-sulfur crude oil is projected to cost about \$79 per barrel in 2010, rising to \$90 in 2015, \$102 in 2020, and \$119 in 2030 (all in 2006 dollars). Oil was over \$130 per barrel in June 2008, but that does not necessarily mean that the AEO numbers are wrong. Other projections are both well above and well below this one.

[2]Carbon dioxide is the main greenhouse gas of concern in this analysis, but other gases, especially methane and nitrous oxide, are emitted as part of the full fuel cycle. These are accounted for with global warming equivalency factors, taken from the literature and the GREET model.

TABLE 6.3 Assumptions in Reference Case

	2005	2020	2035	2050
Total number of LDVs (millions)	220	280	331	369
Share of LDV fleet				
Gasoline ICEVs	99.8%	96.6%	93.2%	91.4%
Gasoline hybrids	0.3%	3.4%	6.8%	8.6%
New LDVs sold per year (millions)	16.2	18.2	20.8	22.4
Share of New LDVs per year				
Gasoline ICEVs	98.6%	94.7%	91.8%	91.1%
Gasoline hybrids	1.4%	5.2%	8.2%	8.9%
Average on-road fuel economy, new gasoline LDVs (mpg)				
Gasoline ICEVs	20.2	29.3	30.6	31.7
Gasoline hybrids	32.1	41.0	42.9	44.5
Gasoline price ($/gallon)	2.32	3.19	3.54	3.96
Vehicle-miles traveled (billion per year)	2,556	3,251	4,243	5,364
Gasoline consumed (billion gallons per year) (includes blends of ethanol up to 10%)	124	132	140	157
Ethanol (billion gallons per year) consumed as:				
Blend in gasoline to 10%	3.4	12.7	15.6	21
E 85	0.01	0.06	0.20	
Greenhouse gas emissions (million tonnes CO_2 equivalent per year)	1,345	1,442	1,527	1,710

were quite adequate for the purpose of scoping the potential growth of HFCVs and their impact.

Modeling Assumptions

- Only U.S. light-duty vehicles (LDVs) are considered. All scenarios assume the same total number of vehicles and vehicle-miles traveled (VMT) as the reference case.
- The AEO high-price case is used as the basis for energy prices.
- The real discount rate is 15 percent (no inflation is included).
- The costs are given in 2005 constant dollars.
- Costs for hydrogen infrastructure technologies (hydrogen production, storage, delivery, refueling stations) draw heavily on DOE's H2A database (DOE, 2007). Both current and 2015 technology numbers are used where available (for production and refueling station technologies).

HYDROGEN SCENARIO ANALYSIS

Reference Case

Cases 1 through 4 (hydrogen and other alternative fueled vehicle) are compared against a reference case. The reference case is based on the high-price case of the AEO (EIA, 2008), which gives projections to 2030 for vehicle miles traveled, vehicle fuel economy, and vehicle fleet composition based on the DOE National Energy Modeling System (NEMS). For input beyond 2030, a vehicle stock model was adapted from the Argonne National Laboratory VISION model to estimate numbers of LDVs, fuel economy, and vehicle energy use to 2050 (Singh et al., 2003; Argonne, 2007). The vehicle stock model keeps track of what vehicles are in the fleet (vehicles are retired after a certain number of years) as increasing numbers of hydrogen vehicles enter the market, allowing the calculation of energy use and greenhouse gas emissions for each year. Table 6.3 and Figures 6.3 to 6.5 summarize the reference case. After the committee had completed its initial analysis, Congress passed the Energy Independence and Security Act of 2007 (EISA), which included a significant increase in fuel economy standards for vehicles. The reference case reflects this new policy.

The reference case includes modest use of biofuels. In 2030, the reference case assumes that 15 billion gallons of ethanol are used per year, 12 billion from corn ethanol and 3 billion from cellulosic ethanol. This gradually increases to 21 billion gallons per year in 2050 (12 billion gallons from corn, 9 billion gallons from cellulosic feedstocks). This is equivalent to about 8-10 percent ethanol by volume in gasoline after about 2020.[3]

[3]EISA includes ambitious goals for biofuels as well as fuel economy. The goal for 2020 is 36 billion gallons, of which 21 billion would be from advanced processes, such as cellulosic ethanol. The committee decided not to include these goals in its reference case for two reasons. First, the likelihood of meeting the biofuel goals appears to be much lower than for fuel

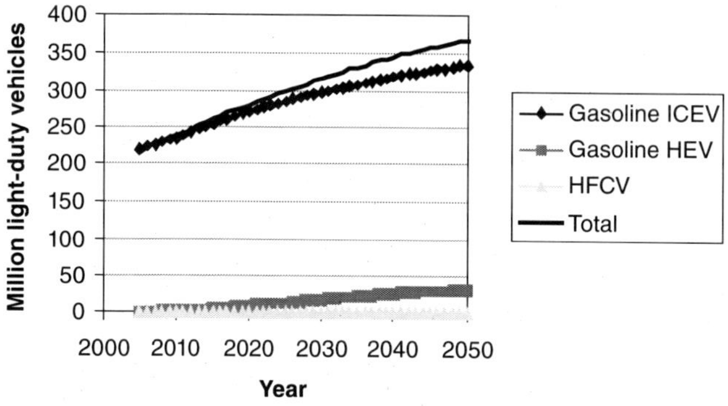

FIGURE 6.3 Reference case: Number of light-duty vehicles in the fleet.

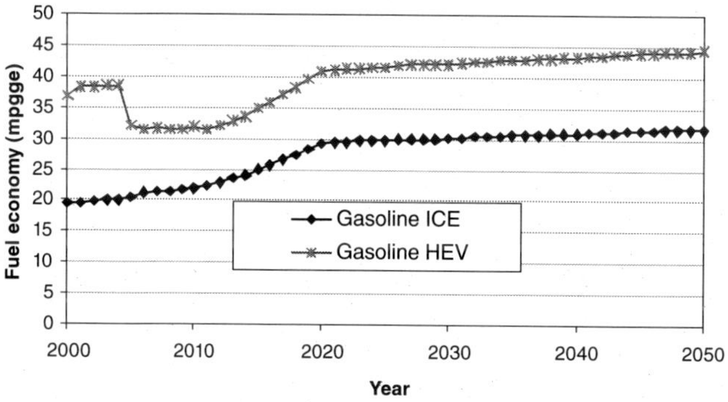

FIGURE 6.4 Reference case: Assumed fuel economies for gasoline ICEVs and gasoline hybrid vehicles (HEVs). (The "dip" in hybrid fuel economy in 2004 occurred when hybrid sport utility vehicles and vans entered the market.)

Hydrogen Cases 1, 1a, 1b

Table 6.1 lists cost and performance assumptions for hydrogen fuel cell vehicles and a gasoline reference vehicle for each hydrogen case. HFCV costs are based on an 80 kW fuel cell "engine" with 5 kg (165 kWh) of compressed hydrogen gas stored on board.[4] (The drivetrain includes the fuel cell and auxiliaries, a hybrid battery, electric motor, wiring, and hydrogen storage.) This is consistent with the following assumptions:

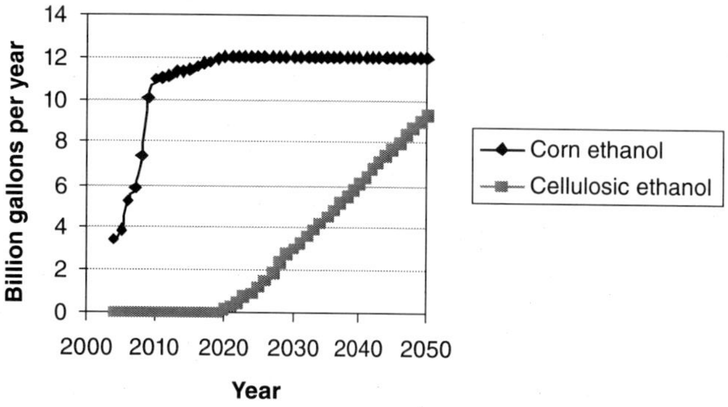

FIGURE 6.5 Reference case: Assumed biofuel use.

- For Cases 1 (Hydrogen Success) and 1a (Accelerated Hydrogen), the fuel cell drivetrain (the fuel cell system, hybrid battery, motor, and auxiliaries) costs the automaker (original equipment manufacturer, or OEM) $50/kW. This corresponds to a fuel cell system cost of $30/kW plus added costs for a hybrid battery, electric motor, and other components. Of the $30/kW fuel cell system cost, about half is due to the fuel cell stack and half to the balance of plant. Hydrogen storage costs the OEM $10/kWh. A model from Kromer and Heywood (2007) shows the total OEM manufacturing cost for drivetrain plus storage to be $71/kW, or a retail price of about $100/kW, giving a drivetrain plus storage price of $7,920.
- For Case 1b (Hydrogen Partial Success), the fuel cell drivetrain costs the OEM $62/kW corresponding to a fuel cell system cost of $50/kW plus added costs for a hybrid battery, electric motor, and other components. Of the $50/kW fuel cell system cost, about 40 percent is due to the fuel cell stack and 60 percent to the balance of plant. Hydrogen storage costs $15/kWh. The total OEM manufacturing cost is $93/kW or a retail price of about $130/kW, giving a drivetrain plus storage price of $10,400.

The drivetrain and fuel storage for a reference gasoline internal combustion engine car are assumed to have an OEM cost of $35/kW plus $300 for the exhaust system. For an 80 kW engine, the OEM cost is $3,100 and the retail price $4,300 ($54/kW). The price for each vehicle is broken down into a drivetrain and a "glider" (the rest of the vehicle). For all vehicles the glider price is the same, $18,750. The HFCV price is assumed to decrease according to a learning curve model developed by Oak Ridge National Laboratory researchers (Greene et al., 2007), based on automobile manufacturers' estimates of fuel cell vehicle costs in mass production (Figure 6.6).

Cases 1 and 1a assume that the HFCV has twice the fuel economy of an efficient gasoline ICEV, described in Case 2 (ICEV Efficiency) below. (The evolving efficient gasoline ICEV in Case 2 has fuel economy of 25.2 miles per gallon

---

economy. Second, the net oil displacement and CO_2 emission reduction are much less certain than for fuel economy improvements. These factors are discussed in Chapter 4.

[4]The 2006 reference gasoline vehicle is based on a midsized five passenger car, with a curb weight of 1,570 kg. As efficiency improves over time, the weight is reduced to about 1,280 kg by 2030. The weight of the corresponding HFCV is 1,320 kg, reflecting heavier components. This reference vehicle is about average for the current new car fleet and is assumed to represent the fleet. Similarly, the HFCV that replaces it is assumed to be representative. HFCVs, like conventional vehicles, will range from small to large, but the fuel savings can still be determined from the average.

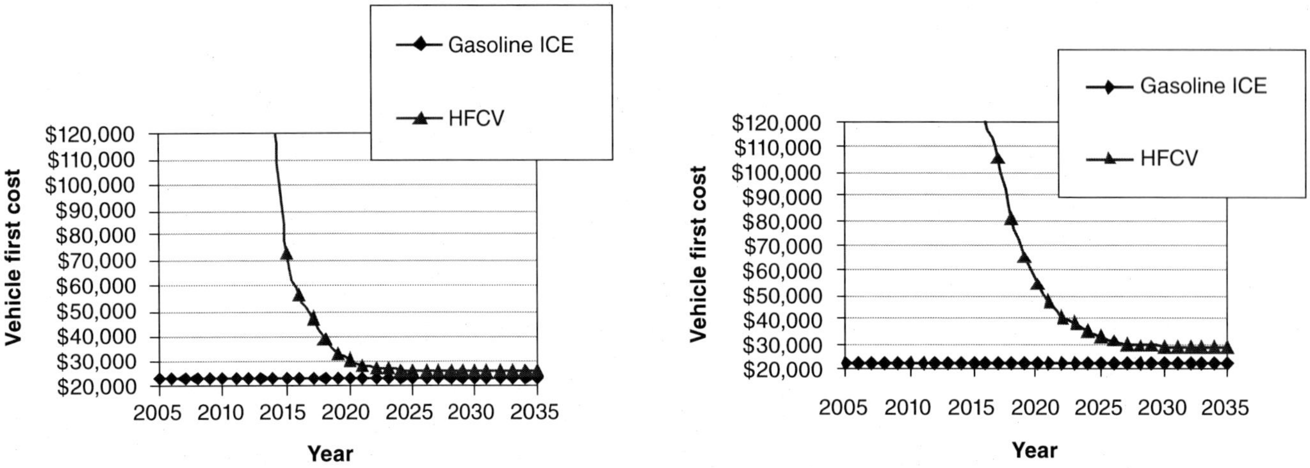

FIGURE 6.6 Assumed retail prices for hydrogen and gasoline vehicles over time for Cases 1 and 1a (left) and Case 1b (right).

(mpg) in 2015 and 42.4 mpg in 2050). The hydrogen fuel cell vehicle has about 1.4 times the fuel economy of the gasoline hybrid in Case 2 (which is assumed to get 36.5 mpg in 2015 and 60.3 mpg in 2050). In Case 1b (Hydrogen Partial Success), the fuel vehicle has 1.75 times the fuel economy of the efficient gasoline ICEV.

Analysis of all three hydrogen cases is detailed in Appendix C. Case 1b (Hydrogen Partial Success) gave rise to only modest reductions in oil use and CO_2 emissions by 2050. The cost of making a transition was roughly twice that of Case 1 (Hydrogen Success) and took several years longer to complete.

Case 1a (Accelerated Hydrogen) gave rise to a marginally faster transition, and the resulting reductions in oil use and CO_2 emissions were 25-33 percent greater than Case 1 (Hydrogen Success) by 2050. However, the estimated transition cost for Case 1a was many times that for Case 1, because it assumed that more of the expensive early vehicles are pushed into the market in the early years of the transition.

For these reasons the committee chose Case 1 (Hydrogen Success) as the maximum practicable case as requested in its statement of task. Cases 1a and 1b are not considered further in this chapter.

Hydrogen Infrastructure Requirements and Costs

The UC Davis SSCHISM steady-state hydrogen supply pathway model (Yang and Ogden, 2007b) is used to design hydrogen infrastructure and estimate delivered hydrogen costs for Case 1 (Hydrogen Success). Hydrogen equipment costs and performance are from the H2A model developed by the Department of Energy (Paster, 2006). The H2A component-level data are combined into complete hydrogen supply pathways from hydrogen production through refueling using the SSCHISM steady-state pathways model developed at the University of California-Davis (Yang and Ogden, 2007b). SSCHISM employs an idealized spatial model of infrastructure layout in cities to design and cost alternative infrastructure pathways. Inputs include information about the level of demand (market fraction of hydrogen vehicles), the city population and size, the number of stations, local feedstock and energy prices, and constraints on viable types of supply. Outputs include the delivered hydrogen cost to the vehicle, hydrogen infrastructure costs, and energy use and CO_2 emissions for different supply pathways. Cost and performance data about hydrogen production and delivery technologies are discussed in Chapter 3.

The committee makes several assumptions in designing the hydrogen infrastructure.

1. *Phased introduction.* There is a phased introduction of hydrogen vehicles and stations in selected large cities, beginning with cities such as Los Angeles and New York (with interest and motivation to implement hydrogen) and moving to other cities over time. This so-called lighthouse concept reduces infrastructure costs by concentrating development in relatively few key areas termed "lighthouse cities." A possible schedule for phased introduction of hydrogen vehicles in various U.S. cities is shown in Figure 6.7. The list of 27 cities was chosen based on hydrogen scenario development work by DOE (Gronich, 2007; Melendez, 2006).

2. *Station "coverage."* Initially, when hydrogen is introduced in each lighthouse city, some minimum number of hydrogen stations is needed to ensure adequate coverage and consumer convenience. This constraint is imposed to help deal with the "chicken-and-egg" problem of assuring hydrogen fuel availability to early non-fleet vehicle owners. This is assumed to be 5 percent of existing gasoline stations in cities (Nicholas et al., 2004; Nicholas and Ogden, 2007). The percentage of hydrogen stations and station capacity over time are shown in Figures 6.8 and 6.9. For the initial introduction of hydrogen vehicles, it is assumed that 100 kg/d stations are

	2012	2013	2014	2015	2016	2017	2018	2019	2020	2021	2022	2023	2024	2025
Los Angeles	1	2	2	25	40	50	85	120	160	190	210	250	270	300
New York, Chicago				25	40	50	85	120	150	175	185	225	240	270
San Francisco, Washington/Baltimore					20	30	55	85	120	140	160	190	210	230
Boston, Philadelphia, Dallas						20	50	85	120	145	165	195	210	220
Detroit, Houston							25	50	80	120	140	160	190	210
Atlanta, Minneapolis, Miami								40	75	100	115	130	160	180
Cleveland, Phoenix, Seattle									45	70	90	120	150	170
Denver, Pittsburgh, Portland, St. Louis, Cincinnati, Indianapolis, Kansas City										60	80	110	130	150
Milwaukee, Charlotte, Orlando, Columbus, Salt Lake City											55	80	110	130
Nashville, Buffalo, Raleigh												40	70	90
Nationwide													260	540

FIGURE 6.7 DOE plan for introduction of light-duty hydrogen vehicles into 27 "lighthouse" cities (thousand vehicles per year introduced between 2012 and 2025). The overall build-up rate corresponds to Case 1. The total number of vehicles in 2025 is 10 million, and 2.5 million vehicles are sold that year. SOURCE: Gronich (2007).

available at 5 percent of gasoline stations for the first several years. (These very early stations might be supplied from the existing industrial hydrogen system, using excess hydrogen from refineries and other industrial or merchant sources.) This is followed by a brief period of building "medium-sized" 500 kg/d on-site steam methane reformers (SMRs) at 5 percent of gasoline stations. As demand grows, capacity is added at each of these stations to make them 1,500 kg/d stations. Beyond about 2022, new 1,500 kg/d stations are added, and the fraction of gasoline stations offering hydrogen increases over time. To account for underutilization of hydrogen station equipment as demand grows, a relatively low system capacity factor of 70 percent is assumed.

The assumed capital costs of different hydrogen production systems are summarized in Table 6.4, based on H2A's future (2015) technology assumptions (DOE, 2007). (See also Chapter 3.)

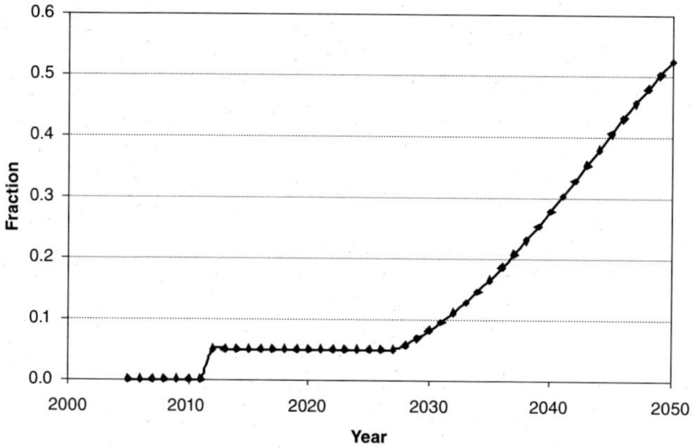

FIGURE 6.8 Fraction of gasoline stations offering hydrogen, 2000-2050.

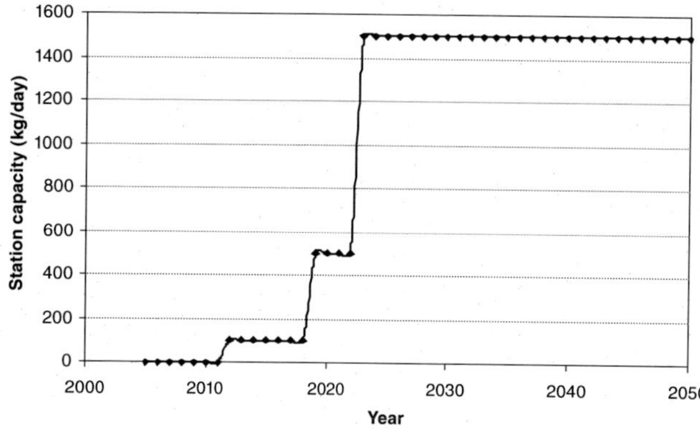

FIGURE 6.9 Capacity of new hydrogen stations by year, 2000-2050.

TABLE 6.4 Assumed Capital Costs for Hydrogen Production Systems

	Plant Size (tonne/d)	FUTURE TECH H2A 2015 Capital Cost (dollars/kg per day)
Central natural gas	50	621
(SMR)	300	400
(production plant only)	400	375
Central coal	250	1,275
(production plant	400	1,170
only)	1,200	950
Central biomass	30	1,260
(production plant	155	860
only)	200	820
On-site SMR (station)	0.1	3,970 ($0.4 million per station)
	0.5	1,811 ($0.9 million per station)
	1.5	1,452 ($2.2 million per station)
On-site electrolysis	0.1	4,325 ($0.4 million per station)
(station)	0.5	2,050 ($1.0 million per station)
	1.5	1,673 ($2.5 million per station)

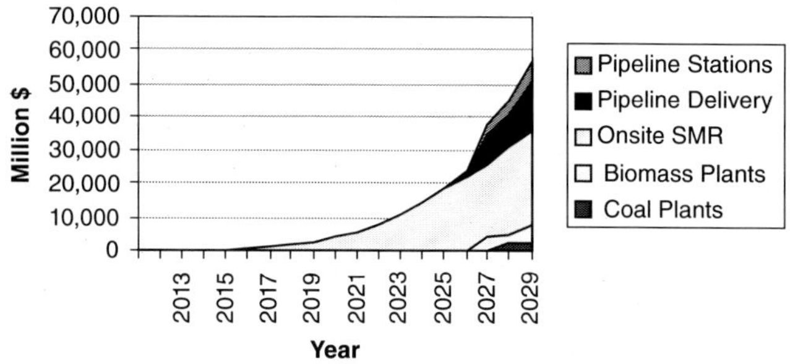

FIGURE 6.10 Early infrastructure capital costs for Case 1.

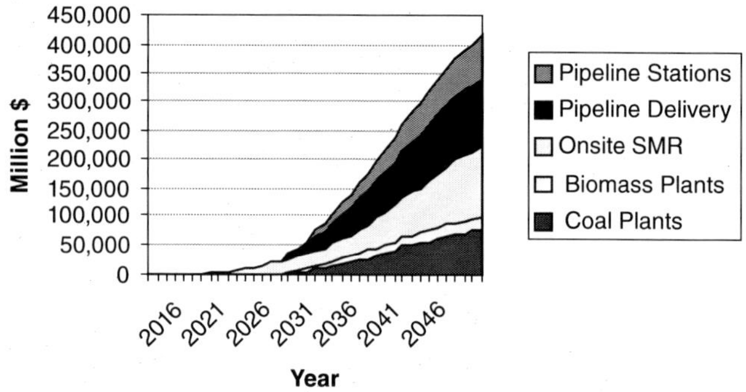

FIGURE 6.11 Capital costs for hydrogen infrastructure.

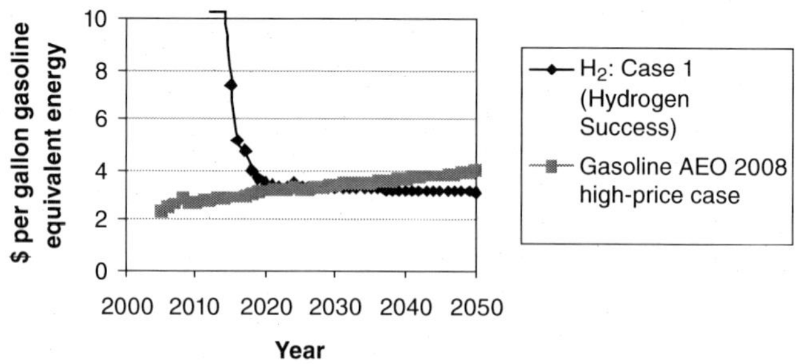

FIGURE 6.12 Estimated average cost of delivered hydrogen in the United States and the assumed gasoline price.

The SSCHISM infrastructure model compares the different possible supply options in Table 6.2 for 73 different U.S. cities, finding the lowest-cost supply in each city at a specified market penetration. The best choice depends on the level of demand, the city size and demand density, and local energy and feedstock prices. For the first 5-10 years, on-site SMRs dominate the hydrogen supply. After that time, central production plants begin to be built in large cities, with truck or pipeline delivery, although on-site SMRs are assumed to persist in smaller cities. All coal hydrogen plants are assumed to have carbon capture and sequestration (CCS). Biomass hydrogen plants are small in size (30-200 tonnes per day) to match the scale of regional biomass supply. This compares to 250-1,200 tonnes per day for coal plants. The analysis uses a regional biomass supply curve (that specifies the amount of biomass available at a certain amount per tonne) (Perlack et al., 2005) to reflect biomass feedstock cost increases as demand grows.

Figure 6.10 shows the capital costs for infrastructure up to 2030. On-site SMRs dominate, with central production and pipeline delivery beginning after about 2027, when the first central production systems using biomass and coal are built. These are accompanied by pipeline delivery systems and stations. Biomass plants appear slightly earlier than coal hydrogen plants, and more of them are built because they are smaller in size. Later, central production dominates in large cities, although on-site reformers persist in other areas (Figure 6.11).

In terms of the amount of hydrogen produced, coal-based hydrogen is the dominant source, with significant contributions from biomass hydrogen as well. However, it is important to note that the delivered costs of hydrogen from coal, biomass, and natural gas central plants are quite close (within $0.50/kg). Thus, the choice of a feedstock may be determined by other factors, such as state policies favoring renewables. The long-term capital cost for infrastructure is roughly 25 percent on-site reformer stations, 25 percent central production plants (most hydrogen comes from coal, with some from biomass), 25 percent delivery systems (pipelines predominate), and 25 percent refueling stations with truck or pipeline delivery. The U.S. average cost of delivered hydrogen is shown in Figure 6.12.

The total infrastructure capital cost is about $2,000 per car served by the system. The total capital costs to build a

TABLE 6.5 Type of Hydrogen Supply over Time

Case 1 (Hydrogen Success)	2020	2035	2050
No. of cars served (percentage of total fleet)	1.8 million (0.7%)	61 million (18%)	219 million (60%)
Infrastructure capital cost	$2.6 billion	$139 billion	$415 billion
Total no. of stations	2,112 (all on-site SMR)	56,000 (40% on-site SMR)	180,000 (44% on-site SMR)
No. of central plants	0	113 (20 coal, 93 biomass)	210 (79 coal, 131 biomass)
Pipeline length (miles)	0	39,000	80,000
Hydrogen demand (tonnes per day)	1,410 (100% NG)	38,000 (22% NG, 42% biomass, 36% coal with CCS)	120,000 (31% NG, 25% biomass, 44% coal with CCS)

NOTE: NG = natural gas.

"steady-state" hydrogen infrastructure to serve the demands in 2020, 2035, and 2050 are estimated in Table 6.5. Note that more than $400 billion is required to build the hydrogen infrastructure to supply the fuel for the HFCV fleet in 2050.

Investment Costs for Hydrogen Fuel Cell Vehicles to Reach Cost Competitiveness

Examining the annual cash flows reveals the total investment required for hydrogen HFCVs to reach "breakeven" with gasoline ICEVs. These are shown in Figure 6.13:

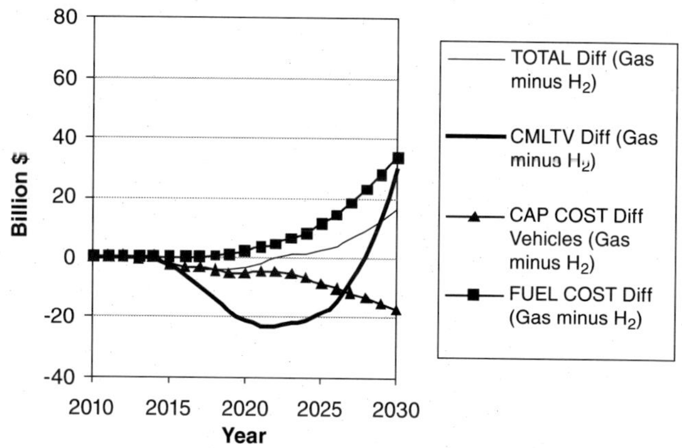

FIGURE 6.13 Cash flows for Case 1.

- The "CAP COST Diff" (dollars per year) is the difference in vehicle price for a gasoline vehicle versus a hydrogen vehicle, summed over all the new HFCVs sold that year. This starts out negative (HFCVs cost a lot more than gasoline vehicles), but small (only a few HFCVs are sold). In the longer term, the annual cost difference continues to grow, as HFCVs are assumed to always cost more than gasoline cars.
- "FUEL COST Diff" (dollars per year) is the annual difference in fuel costs for HFCVs (counting all HFCVs currently in the fleet) compared to what would have been paid to fuel comparable gasoline-fueled vehicles. Hydrogen soon becomes less costly as a fuel on a cents-per-mile basis, so this difference becomes positive around 2017.[5] This analysis assumes that for a new fuel such as hydrogen with a small HFCV fuel tank (approximately 5 to 8 gallons of gasoline equivalent), consumers would value fuel on a cost-per-mile traveled basis (dollars per mile) rather than a cost-per-gallon-equivalent basis, as they do now for gasoline.
- "TOTAL Diff" (dollars per year) is the cash flow, which equals the economy-wide cost per year of pursuing a fuel cell market introduction plan. The cash flow is defined

[5]Hydrogen fuel becomes cost competitive with gasoline (on a cents-per-mile basis) in about 2017, when hydrogen costs are still fairly high, about $5.60/kg. This is because the hydrogen vehicle is assumed to have a fuel economy 2.0 times greater than the gasoline vehicle, and the gasoline price in the AEO high-oil-price case is $2.80 per gallon. This analysis compares the cost of hydrogen with the price of gasoline. The committee decided this would be the most straightforward comparison because it would be difficult to estimate a price for hydrogen without a model for all its uses in the economy, and it is hard to estimate the cost of gasoline, which depends on many complex factors. Gasoline prices include federal, state, and local taxes. One could argue that hydrogen should be competed against the untaxed gasoline price. However, other alternative fuels such as ethanol are untaxed to encourage their adoption, and the committee decided to give hydrogen the same advantage. It should be noted that much of the revenue raised by gasoline taxes goes to highway maintenance and other necessary functions that continue no matter what type of vehicles travel on them. As discussed in Chapter 7, these revenues will have to be replaced from other sources if alternative fuels remain untaxed. On the other hand, the price of gasoline does not include the cost of externalities that hydrogen is intended to address: CO_2 emissions and oil imports.

as the gasoline vehicle minus the annual cost of the hydrogen vehicle. This starts out negative, but becomes positive in 2023, when the cost of hydrogen vehicles becomes less than that of a similar number of gasoline vehicles. While the HFCVs first cost remains higher than for the gasoline car, the net fuel cost savings make the annual cash flow positive at the breakeven year.

- "CMLTV Diff" (dollars) is the cumulative cost difference, the sum of the cost difference over time (starting in 2012), providing a yearly tally of the total funds that would have to be invested to make HFCVs competitive. At first, there is a negative cash flow (early HFCVs cost more than gasoline cars), but eventually as HFCV and hydrogen fuel costs come down, the negative cash flow "bottoms out" in 2023 at a minimum of about $22 billion, when about 5.6 million fuel cell vehicles have been produced. This minimum is the "buydown" investment that must be supplied to bring the HFCV to cost competitiveness.

Most of the negative cash flow is due to the high price of the first few million fuel cell vehicles. This is not surprising, since, initially, fuel cell vehicles cost a lot more than gasoline vehicles (see Figure 6.6). The subsidy that might be needed by automakers or buyers is the sum of the difference in costs between HFCVs and gasoline cars, each year between vehicle introduction in 2012 and life-cycle cost (LCC) breakeven in 2023. This ***cumulative*** difference in vehicle first cost for HFCVs (as compared to a reference gasoline vehicle) is about $40 billion (averaged over the 2012-2023 buydown period, this is about $7,000 per car, or an average of $3.3 billion per year for 12 years). Transition dates and costs are summarized in Table 6.6, relative to a reference gasoline vehicle.

The buydown cost is quite sensitive to assumptions for key factors. For example, if fuel cell vehicles could be introduced at their "learned-out cost" (e.g., the cost of HFCVs once they have become technically mature and are manufactured at large scale), buydown cost requirements for vehicles would be greatly reduced, and fuel cell vehicles would become competitive almost immediately. In this case, the primary transition cost would be building a hydrogen infrastructure to the point at which hydrogen is competitive as a fuel (fuel cost per kilometer), on the order of a billion dollars. Note that this would happen much sooner than the vehicles reaching cash flow breakeven (see bottom row in Table 6.6). Box 6.1 explores the sensitivity of the results to assumptions on HFCV fuel economy and incremental costs, and the cost of hydrogen and gasoline.

TABLE 6.6 Transition Costs and Timing for Hydrogen Cases

Breakeven Year (annual cash flow > 0)	2023
Cumulative life-cycle cost difference (between HFCV and gasoline reference car) to breakeven year	$22 billion
Cumulative vehicles first-cost difference (between HFCV and gasoline reference car) to breakeven year	$40 billion (~$3.3 billion/yr)
Number of HFCVs at breakeven year (millions)	5.6 (1.9% of fleet)
Hydrogen cost at breakeven year	$3.3/kg
Hydrogen demand; number of hydrogen stations at LCC breakeven year	4,200 tonnes/d; 3,600 stations
Total cost to build infrastructure for demand at LCC breakeven year	$8.2 billion
Year when hydrogen fuel cost per kilometer = gasoline price per kilometer	2016
Hydrogen cost ($/kg)	5.20
Gasoline price ($/gal)	2.70
Total cost to build infrastructure to meet demand in 2023 (LCC breakeven year)	$0.5 billion (1,000 small on-site SMR stations)

RESULTS: COMPARISON OF GREENHOUSE GAS EMISSIONS AND OIL DISPLACEMENT FOR SCENARIOS

Assumed Greenhouse Gas Emissions for Fuels

Until 2020, all hydrogen comes from on-site SMRs with a CO_2 release of 100 g CO_2 equivalent per megajoule of fuel. After that time, low-carbon sources such as biomass hydrogen and hydrogen from coal with carbon capture and storage are phased in. By 2050, roughly 31 percent of hydrogen is produced via on-site SMRs, the remainder via low-carbon sources (44 percent coal with CCS, 25 percent biomass H_2). Thus, the overall emissions for hydrogen supply in 2050 are 37 g CO_2/MJ fuel, based on the CO_2 values given in Table 6.7, which shows the assumptions regarding the well-to-wheels emissions associated with different fuel supply pathways. In all cases, the carbon emissions from the hydrogen supply are assumed to follow the curve shown in Figure 6.14, where CO_2 emissions are shown as declining linearly between 2020 and 2050. The average CO_2 emissions might fall faster than this, because most new capacity after

TABLE 6.7 Assumed Greenhouse Gas Emissions per Unit of Fuel Consumed

Fuel	Greenhouse Gas Content of Fuels ($g\ CO_2$ equivalent/MJ fuel used LHV basis)
Conventional gasoline	90 (AEO 2007 projections to 2030 show this staying approximately constant)
Hydrogen from on-site reformation	100 (TIAX, 2007)
Hydrogen from coal with CCS, pipeline delivery	16 (92% CO_2 capture rate, assuming U.S. electricity mix) 0 (100% CO_2 capture rate, assuming zero-carbon electricity is used for CO_2 compression) (adapted from Ruether et al., 2005). Coal mining, transport, and plant and mine construction emissions are estimated to be about 3.8 g CO_2 eq/MJ H_2, but this is counterbalanced by GHG savings of about 4.3 g CO_2 eq/MJ, due to exported electricity from the plant)
Hydrogen from biomass, pipeline delivery	10 (TIAX, 2007)
Hydrogen from electrolysis using zero-carbon electricity (wind, hydro, solar)	0
Ethanol from corn	70 (22% reduction relative to gasoline)
Ethanol from cellulose	13 (85% reduction relative to gasoline)

SOURCES: Ruether et al. (2005); TIAX (2007).

2025 is likely to be very low carbon in these scenarios (70 percent biomass hydrogen or coal hydrogen with CCS).

Case 1 (Hydrogen Success)

Oil Displacement

Figure 6.15 estimates gasoline consumption for the Hydrogen Success case and the reference case. Oil displacement is about 0.8 percent in 2020, rising to 24 percent in 2035 and 69 percent in 2050. Although it takes several decades for hydrogen's impact to be seen, beyond 2025 it enables growing reductions in both greenhouse gas emissions per year and annual oil use. Hydrogen may be important to achieve long-term stabilization goals requiring deep cuts in carbon or oil use. Figures 6.15 and 6.16 show a dip in the reference case after about 2020. This occurs even though the increase in fuel economy of new cars levels off because the entire on-road fleet fuel economy increases as new efficient vehicles replace older vehicles.

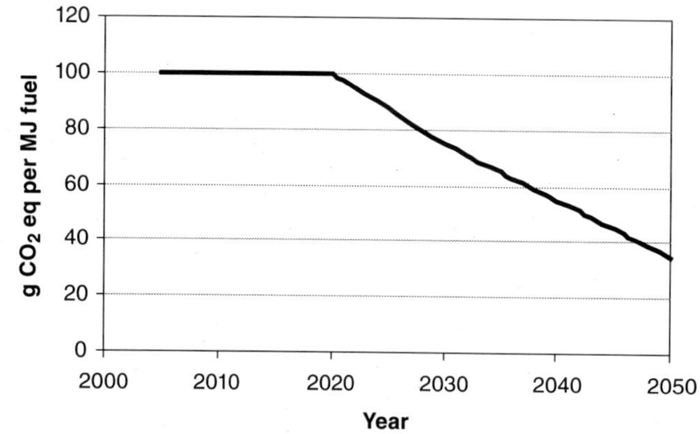

FIGURE 6.14 Greenhouse gas emissions from hydrogen supply over time.

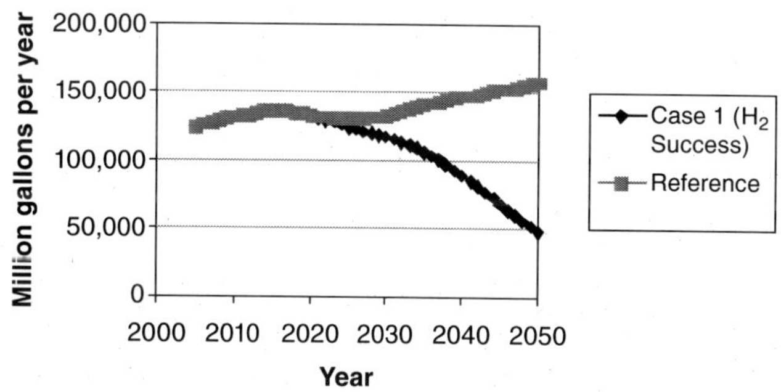

FIGURE 6.15 Case 1 gasoline consumption relative to the reference case.

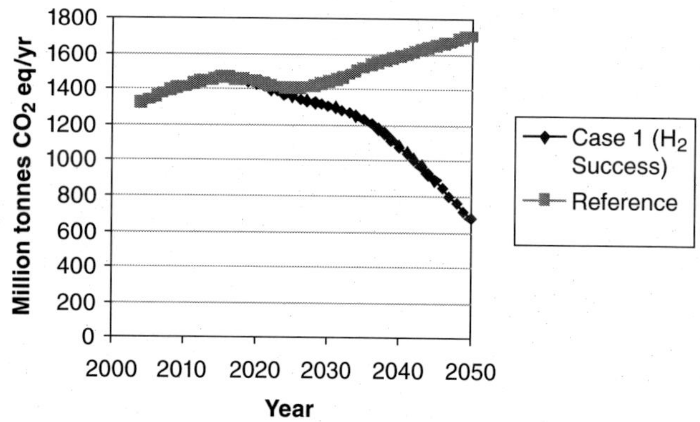

FIGURE 6.16 Case 1 greenhouse gas emissions relative to the reference case.

Greenhouse Gas Reductions

Figure 6.16 shows the committee's estimate of the reductions in greenhouse gas emissions for Case 1 (Hydrogen Success) relative to a reference case, where no hydrogen technologies are introduced. Greenhouse gas emissions per

BOX 6.1
Sensitivity of Breakeven Analysis Results to Changes in Assumptions

The results of an analysis of the costs and timing of a hydrogen transition depend on many assumptions and inputs and are sensitive to changes in four important input parameters:

1. The fuel economy of an HFCV compared to a reference gasoline vehicle;
2. The incremental cost of an H2FCV compared to a gasoline reference vehicle;
3. The cost of hydrogen; and
4. The cost of gasoline (dollars per gallon).

The values of these parameters for Case 1 (Hydrogen Success) are shown in Table 6.1.1. Also shown are potential high and low values for each parameter. Each parameter is varied over this range.

Three key metrics that describe a hydrogen transition are:

1. The "breakeven year";
2. The "breakeven cost" (e.g., cumulative cash flow to get to the breakeven year); and
3. The total capital cost (the incremental cost for HFCVs + the infrastructure cost) to get to the breakeven year.

TABLE 6.1.1 Range Over Which Parameter Values Can Vary for Case 1

Parameter	Low Value	Case 1 Value	High Value
Fuel economy of HFCV versus fuel economy of efficient gasoline car	1.3	2	3
Incremental cost of HFCV compared to reference gasoline vehicle	1,713 (FC system = \$50/kW; H_2 storage = \$2/kWh)	3,600 (FC system = \$50/kW; H_2 storage = \$10/kWh)	6,800 (FC system = \$62/kW; H_2 storage = \$18/kWh)
Incremental cost of H_2 compared to Case 1 long-term cost	–\$2/kg	0	\$2/kg
Price of oil	0.5 × high-price case (oil at \$40-\$60/bbl in 2012-2030)	Oil price from AEO 2008 high-price case (oil at \$80-\$120/bbl in transition period 2012-2030)	1.3 × high price case (oil at \$105-\$160/bbl in 2030)

NOTE: bbl = barrel; FC = fuel cell.

The value for each of these three metrics is shown as a function of the variables in a series of three "spider plots" (Figures 6.1.1-6.1.3). In each plot, the *x*-axis is the ratio of the variable to its Case 1 value (shown in Table 6.1.1.). (The normal Case 1 input is represented by a value of "1" on the *x*-axis. So, an HFCV incremental cost of \$3,600 corresponds to an *x*-value of 1, while an HFCV incremental cost of \$6,800 corresponds to an *x*-value of 6,800/3,600 = 1.88.) When the input parameter is varied from its low to its high values, the value of *x* changes from 0.5 to 2. The metric varies as the input changes. This shows the sensitivity of the results to changes in the input variables.

As expected, the breakeven year is delayed and buydown costs are higher if the HFCV price is higher, the HFCV is less efficient, or hydrogen costs more than in Case 1. Breakeven occurs faster and the buydown cost is less for higher oil prices.

For example, if hydrogen costs \$1/kg more than expected, there is relatively little impact on the breakeven year or the transition cost. However, if the cost of hydrogen is \$2/kg higher than expected, this delays breakeven by 12 years (from 2023 to 2035) and raises the breakeven cost by almost a factor of three (from about \$23 billion to \$61 billion). If the oil price is 1.3 times the AEO's projected high-price case (e.g., about \$100 to \$160 per barrel of oil in the transition period between 2012-2030), the breakeven year is accelerated slightly. However, if oil prices drop to 70 percent of the AEO projections (e.g., about \$55 to \$85 per barrel of oil during the transition period 2012-2030), breakeven is delayed 5 years (from 2023 to 2028) and the buydown cost rises from \$23 billion to \$31 billion. If oil prices drop to 50 percent of the AEO high-price case (\$40 to \$60 per barrel during the transition), breakeven is delayed further to 2035, and the buydown cost almost triples to \$61 billion.

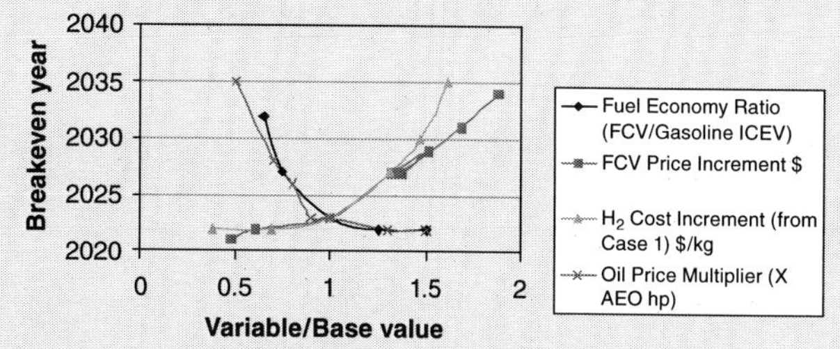

FIGURE 6.1.1 Sensitivity of breakeven year to changes in HCFV fuel economy, HFCV price, H_2 cost, and gasoline price.

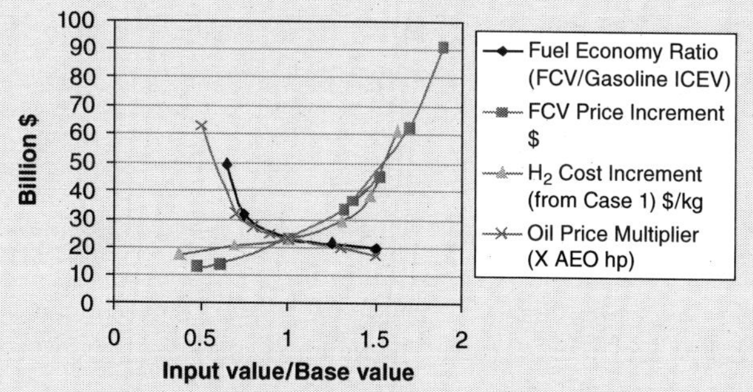

FIGURE 6.1.2 Sensitivity of buydown cost (billion dollars) to changes in HFCV fuel economy, HFCV price, H_2 cost, and gasoline price.

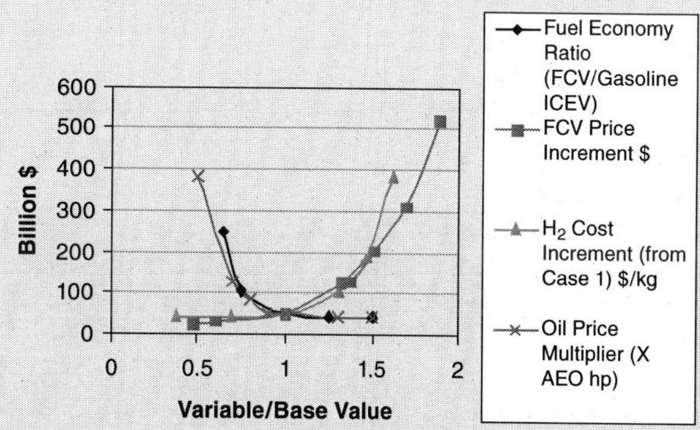

FIGURE 6.1.3 Sensitivity of capital investment to breakeven year (incremental price of HFCVs + H_2 infrastructure capital, billion dollars).

year are reduced by about 0.7 percent in 2020, 19 percent in 2035, and 60 percent in 2050 compared to the reference case.

Case 2 (ICEV Efficiency)

Case 2 is an "evolutionary," not revolutionary, scenario. The committee assumes that currently available improvements in gasoline internal combustion engine technology are used to improve fuel economy (rather than power and acceleration).

A range of more efficient advanced gasoline technologies could be implemented in 2010-2035 as described in Chapter 4. In this scenario, a new "high-fuel-economy" gasoline vehicle is introduced, as well as a hybrid gasoline vehicle, and these capture growing market share over time. By 2035 (2050), 42 percent (85 percent) of new LDVs and 30 percent (60 percent) of the fleet are gasoline hybrids, and the remaining non-hybrid cars have high fuel economy. This is shown in Figure 6.17.

In 2010, the new gasoline vehicle is assumed to have an on-road fuel economy of 22.2 mpg, the hybrid 31.9 mpg. (These values are selected to match the reference case up to 2010.) The fuel economy of each vehicle is then assumed to improve as follows and discussed in Chapter 4:

- 2.6 percent per year from 2010 to 2025,
- 1.7 percent per year from 2026 to 2035, and
- 0.5 percent per year from 2035 to 2050.

The on-road new car fuel economy over time is plotted in Figure 6.18. Note that this is similar to the reference case in Figure 6.4 up to about 2020. Beyond this, Case 2 is significantly more efficient; by 2050, gasoline ICEV and hybrid cars are about 35 percent more efficient than in the reference case, which incorporates the new CAFE standards.

The committee did not project increased market share for diesel engines in this scenario because of the uncertainty over the costs of meeting future tailpipe emission specifications and consumer acceptance, considering the poor history of diesels in U.S. automobiles. However, advanced diesel power

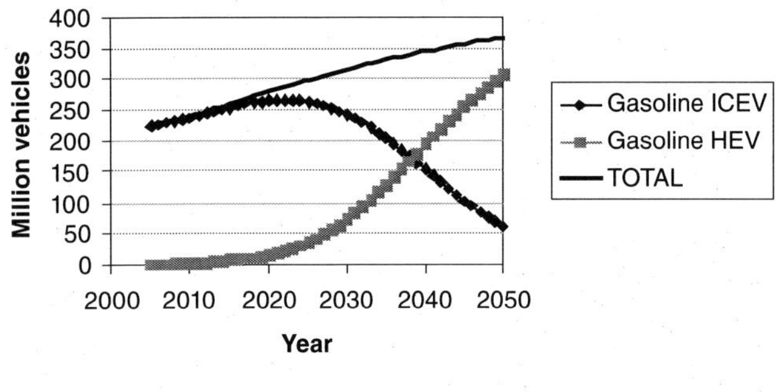

FIGURE 6.17 Case 2 assumed market penetration for gasoline ICEVs and advanced gasoline HEVs.

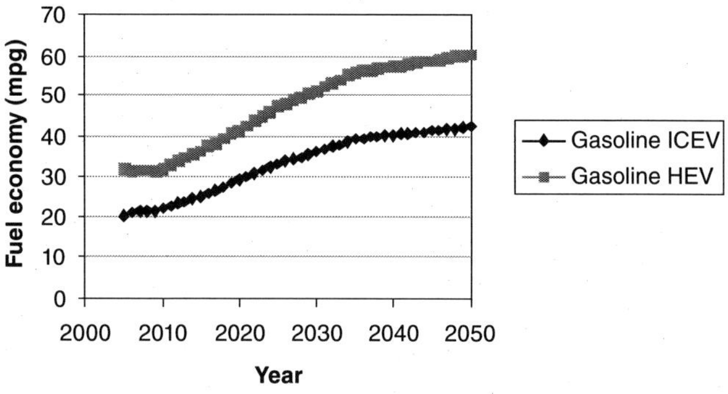

FIGURE 6.18 Case 2 assumed on-road fuel economy for new gasoline ICEVs and gasoline hybrid ICEVs over time.

trains could offer an additional 15 percent reduction in fuel consumption and CO_2 emissions over advanced conventional spark ignition power trains and have cost advantages over hybrid electric vehicles (see, for example, Adrian, 2004). In a high-fuel-cost environment, they could become a growing fraction of LDV sales with the some shifts in government positions on diesels and a positive public relations program. Thus, to the extent that diesels can penetrate the market, this scenario may understate potential fuel savings.

The same vehicle stock model used in the reference case keeps track of the vehicle numbers and vintages of advanced gasoline cars and gasoline hybrids on the road in any year. This allows calculation of oil consumption and greenhouse gas emissions for each year.

Oil Displacement

Gasoline consumption for the case above is estimated in Figure 6.19. Improving fuel economy is a very effective way to cut gasoline use. Gasoline consumption in 2020 is only slightly reduced relative to the reference case, which includes rapidly improving fuel economy, but in 2035 it is down by 35 billion gallons per year (25 percent), and in 2050, by 64 billion gallons per year (40 percent).

Greenhouse Gas Reductions

Greenhouse gas emissions show a similar trend (Figure 6.20). Fuel economy improvements can yield increasing reductions in greenhouse gases. Greenhouse gas emissions are reduced by about 24 million tonnes of CO_2 equivalent per year (1.7 percent) by 2020, 385 million tonnes (25 percent) by 2035, and 700 million tonnes (41 percent) by 2050.

Based on projected gasoline prices and cost estimates for improved fuel economy, it appears that gasoline hybrids and advanced gasoline vehicles would pay for themselves on a life-cycle cost basis, so no external subsidy should be needed. A simple calculation shows that increasing fuel economy from 30 to 45 mpg in a car that travels 15,000

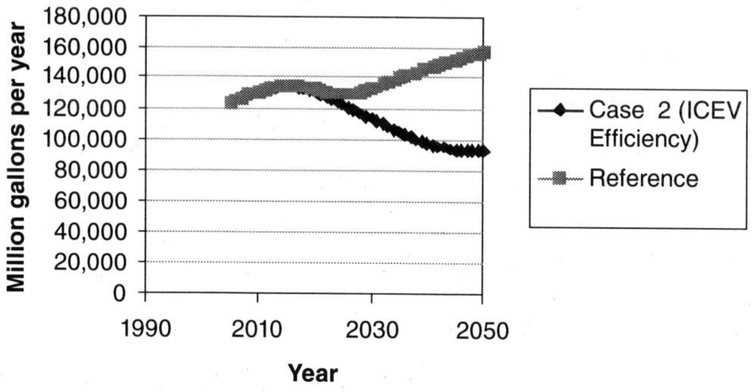

FIGURE 6.19 Gasoline consumption for Case 2 and for the reference case.

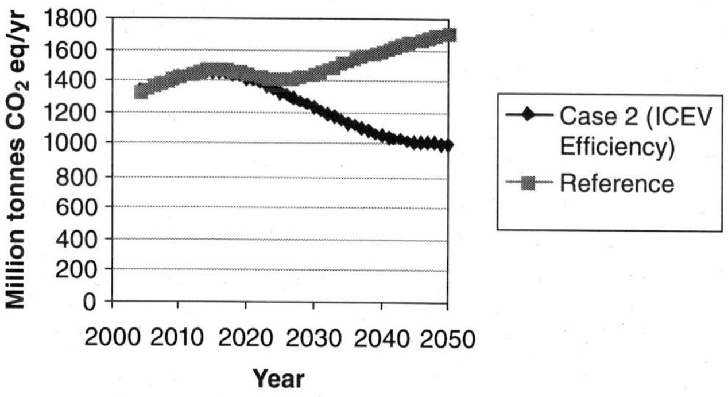

FIGURE 6.20 Greenhouse gas emissions for Case 2 and for the reference case.

miles a year saves 167 gallons of gasoline per year. If a car is kept 10 years and gasoline costs $2.5 per gallon, the present value of fuel savings amounts to $2,565 (assuming a 10 percent discount rate). This would be enough to pay for the difference in first cost between a conventional gasoline car and a hybrid, which is estimated to be $1,800-$2,500 per car (Kromer and Heywood, 2007).

Case 3: Biofuels

Case 3, Biofuels, considers an emphasis on biofuels in the transportation sector. Assumptions about the introduction of biofuels are summarized in Table 6.8. Annual production levels for various biofuels are plotted in Figure 6.21. The detailed assumptions are discussed in Chapter 4. For reference, the maximum practicable case, in the committee's judgment, is 700 million dry tons of sustainable biomass available in 2050, with all of the biomass used for cellulosic ethanol production at 90 gallons per dry ton. This would total 63 billion gallons of ethanol per year in 2050, the amount assumed in Case 3.

The AEO 2007 reference case includes about 12 billion gallons of corn ethanol by 2030 plus an additional 3 billion gallons of cellulosic ethanol. The committee extended this to assume that in the reference case, cellulosic ethanol production reaches 9 billion gallons per year by 2050. The assumed biofuels use in the reference case is shown in Figure 6.5.

TABLE 6.8 Assumed Biofuel Use in Case 3

	Corn Ethanol	Cellulosic Ethanol	Biobutanol	Biodiesel	F-T Diesel via Biomass Gasification
Production (billion gallons)					Timing of introduction similar to that for biomass H_2, toward end of scenario time frame
• 2002					
• 2006	2.5				
• 2008	6			0.25	
• 2010	8	1			
• 2015	10	6		0.6 (2012)	
• 2020	12 (max = 30% of expanded corn crop)	16	0.1		
• 2025		28	0.6	1.5 (max = 30% of soy crop, limited by land)	
• 2030		36	2.6		
• 2035	12	44	4.0		
• 2050		63			
CO_2 reduction (energy equivalent basis)	22% relative to gasoline	85% relative to gasoline		80% relative to regular diesel	
Oil reduction (energy equivalent basis)[a]	1 gallon gasoline equivalent ethanol replaces 0.96 gallon oil	1 gallon gasoline equivalent ethanol replaces 0.93 gallon oil		1 gallon biodiesel replaces 0.95 gallon oil	

[a]Gasoline = 119,000 Btu/gal, ethanol = 80,000 Btu/gal, and biodiesel is equivalent to regular diesel.

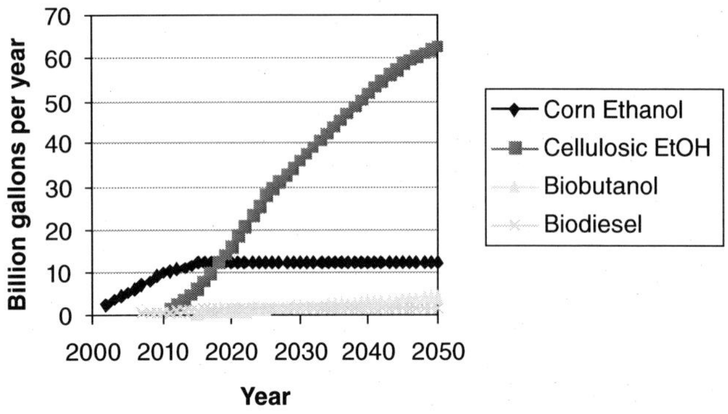

FIGURE 6.21 Annual production of biofuels assumed for Case 3.

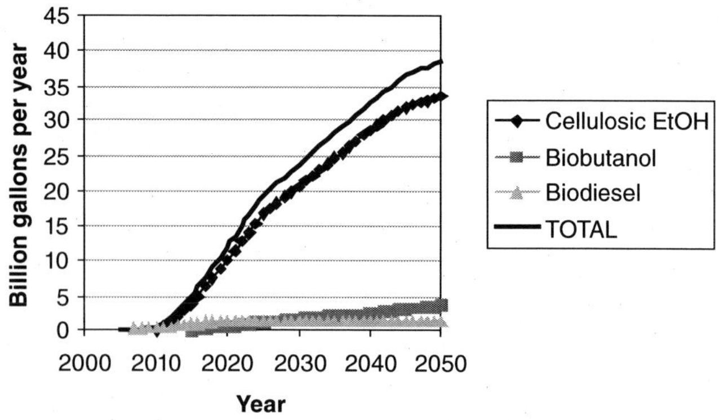

FIGURE 6.23 Case 3: Oil displacement relative to the reference case.

The reference case corresponds to about 10 percent ethanol by volume in gasoline by 2050. In Case 3 (Biofuels), this is expanded to include an additional 54 billion gallons of biofuels per year (beyond the reference case). This is roughly four times the biofuel use in the reference case or about 40 percent ethanol in gasoline plus some other advanced biofuels. ***It is important to note that the greenhouse gas and oil reductions shown for Case 3 are relative to a reference case that already includes biofuel use up to 10 percent of gasoline by volume.***

In the Biofuels case, the assumed corn ethanol use is the same as in the reference case. The difference is that more cellulosic and other advanced biofuels are produced. The additional biofuels production assumed in the Biofuels case (compared to the reference case) is shown in Figure 6.22.

Gasoline displacement for Case 3 is shown in Figure 6.23. The total is about 12 billion gallons per year by 2020 and 39 billion gallons per year in 2050.

Greenhouse gas emissions reductions are given in Figure 6.24 for the Biofuels case. The total reduction in greenhouse gas emissions is about 8 percent from the reference case by 2020, rising to 23 percent by 2050. The committee has not estimated the costs of building biofuel production plants or changes in the fuel distribution infrastructure that might be needed.

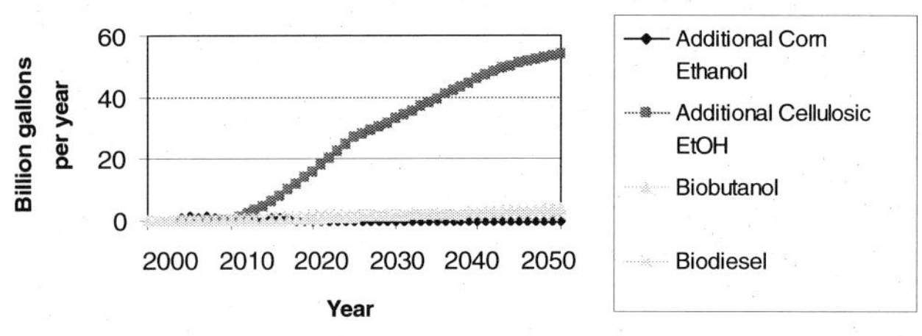

FIGURE 6.22 Case 3: Added biofuel production relative to the reference case.

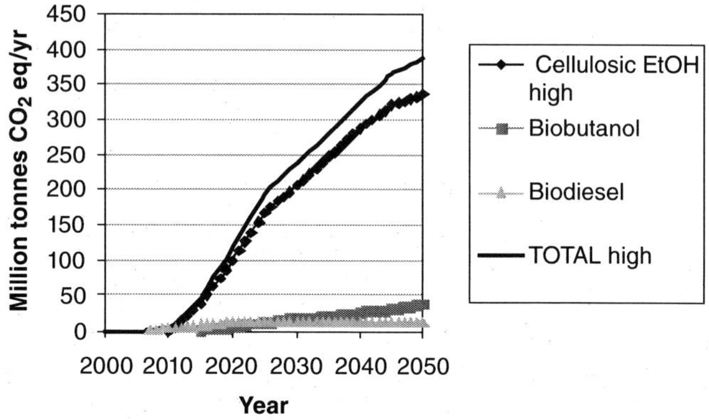

FIGURE 6.24 Case 3: Greenhouse gas emission reductions relative to the reference case.

Comparison of Scenarios

The estimated savings in gasoline use and greenhouse gas emissions for each case are plotted in Figures 6.25 and 6.26. In the near to mid term, improving the fuel economy of gasoline vehicles will be the most effective option for reducing oil use and greenhouse gas emissions. This is already incorporated in the reference case up to 2020, but continued improvements thereafter could match the savings from hydrogen until about 2035. Biofuels could begin to make a difference sooner than hydrogen, which takes more time to implement, assuming cellulosic ethanol comes online in 2010. After about 2032, however, Case 1 (Hydrogen Success) would lead to greater greenhouse gas reductions per year than Case 3 (Biofuels). By 2040, the Hydrogen Success scenario offers about twice the greenhouse gas reduction and oil savings per year as the Biofuels scenario, and by 2050, almost three times the reduction. This clearly illustrates the time frames for different technologies and the total contributions they might make by 2020 and beyond. Although efficiency and biofuels could contribute sooner, hydrogen would surpass the annual savings achievable with either after

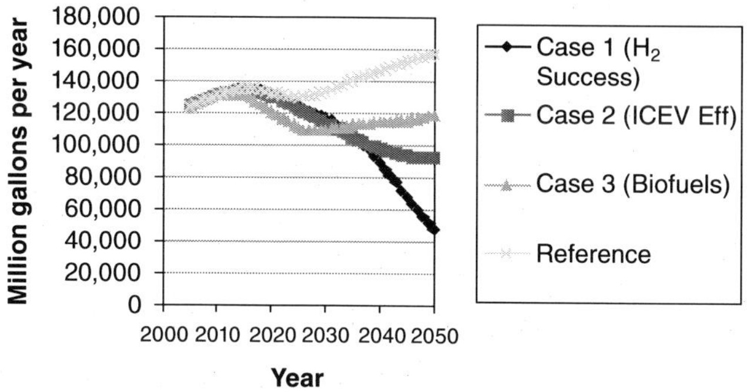

FIGURE 6.25 Oil consumption for Cases 1-3 compared.

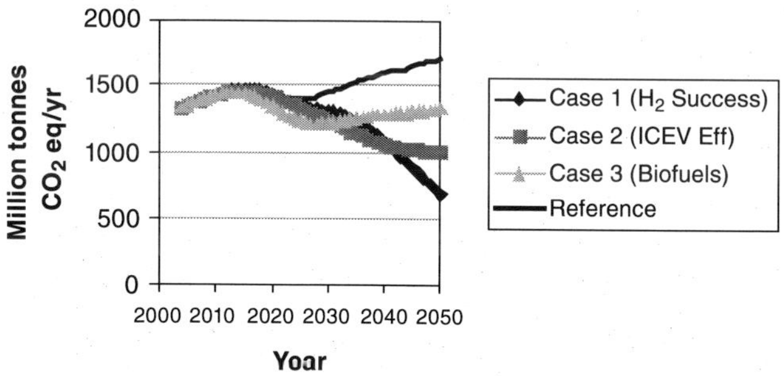

FIGURE 6.26 Greenhouse gas emissions for Cases 1-3 compared.

2035. This result highlights the long time constants inherent in changing the energy system, as well as the need to develop long-term, very low carbon options.

COMBINED APPROACHES TO REDUCING GREENHOUSE GAS EMISSIONS AND OIL USE

The previous section looked at the potential impact of implementing one technological measure at a time. In the future, reducing oil use and greenhouse gas emissions probably will be increasingly important, and it is likely that a combination of approaches would be implemented. However the reductions for each approach cannot simply be added because they affect each other. This section considers several combinations.

Case 1 + Case 2: HFCVs + ICEV Efficiency

This case combines higher gasoline vehicle efficiency with introduction of hydrogen fuel cell vehicles (Case 1 and Case 2). The results are shown in Figures 6.27 and 6.28. Comparing Figure 6.27 and 6.25, gasoline consumption in 2035 is about 18 billion gallons per year lower for the combined case than for HFCVs alone. By 2020, efficiency reduces greenhouse gas emissions by about 1.8 percent relative to the base case. Beyond 2030, HFCVs lead to deeper cuts in emissions than would be possible with efficiency alone.

Case 3 + Case 2: Biofuels + ICEV Efficiency

Combining higher gasoline vehicle efficiency with biofuels yields much greater reductions in oil use and greenhouse gas emissions than are possible with biofuels alone. This is shown in Figures 6.29 and 6.30, which combine Cases 2 and 3. By 2020, biofuel use alone could reduce annual oil use by about 8 percent, with efficiency bringing the total to 10 percent. In the longer term, the effect of efficiency improvements dominates, with biofuels saving 23 percent of gasoline use or greenhouse gas emissions in 2050 and efficiency an additional 41 percent. This strategy "stretches" limited biomass resources to fuel more vehicle miles traveled per acre of land.

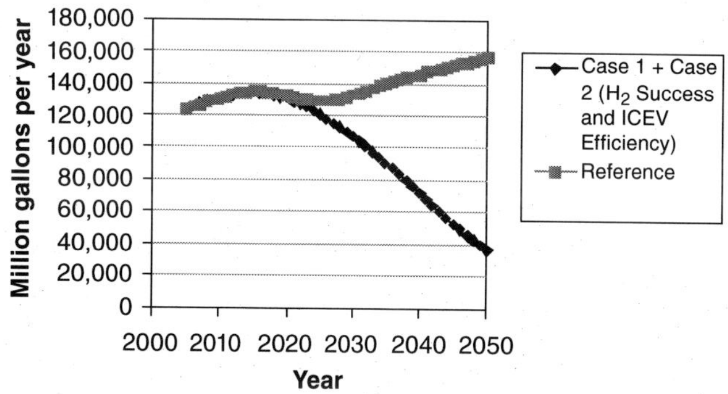

FIGURE 6.27 Oil use for Cases 1 and 2 combined.

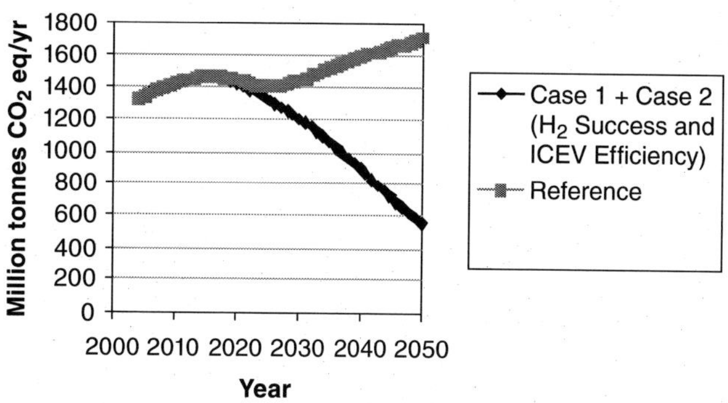

FIGURE 6.28 Greenhouse gas emissions with HFCVs for Cases 1 and 2 combined.

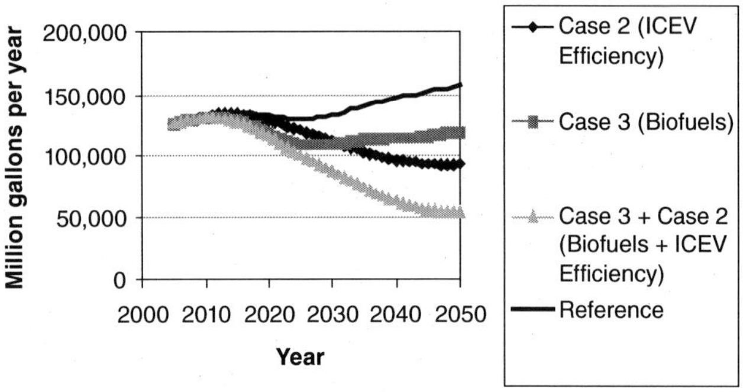

FIGURE 6.29 Oil use for Cases 2 and 3 combined.

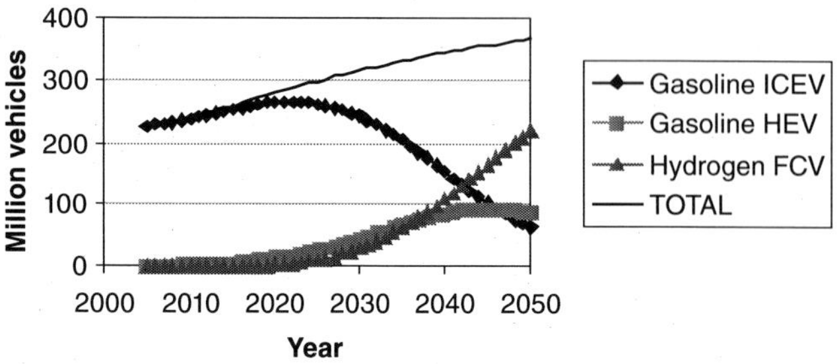

FIGURE 6.31 Assumed number of vehicles in the fleet for Case 4.

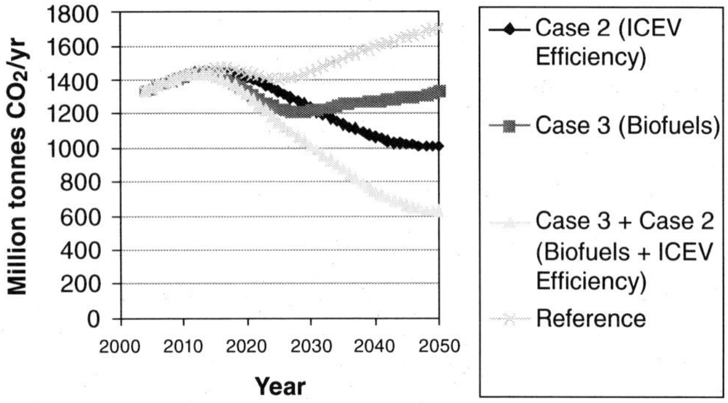

FIGURE 6.30 Greenhouse gas emission reductions for Cases 2 and 3 combined.

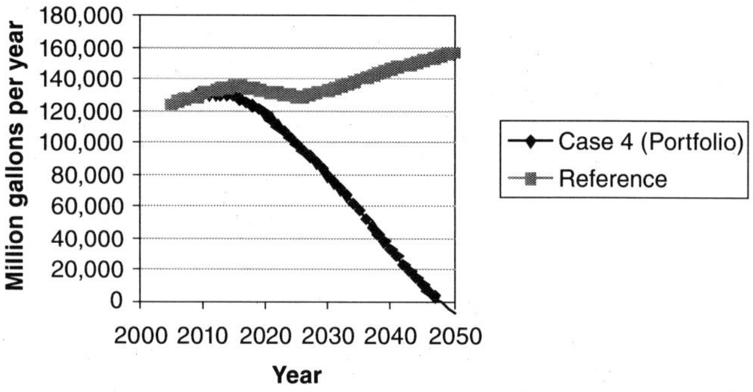

FIGURE 6.32 Oil use in million gallons per year for Case 4.

Case 4 (Portfolio): Implement Efficient ICEVS plus Biofuels and Hydrogen FCVs

Case 4 combines all three of the major options discussed above. Figure 6.31 shows the assumed numbers of vehicles in the fleet over time. Note that the number of hydrogen vehicles is the same as in Case 1 (Hydrogen Success) and the number of gasoline ICEVs is the same as in Case 2 (ICEV Efficiency). The number of advanced gasoline ICEVs (hybrids) increases, but eventually loses market share to HFCVs.

Figure 6.32 shows the reduction in petroleum consumption over time for Case 4. Gasoline use is virtually eliminated in the light duty vehicle fleet by 2050. Table 6.9 shows the reduction in gasoline use for the four cases relative to the reference case.

Table 6.10 lists the emissions reductions relative to the reference case for the four cases. Emissions of greenhouse gas over time are shown in Figure 6.33. The cumulative impact of reductions is shown in Figure 6.34. With a combined approach including efficiency, biofuels, and hydrogen fuel cells, it is possible to reduce CO_2 emissions by about 90 percent and gasoline use by 99 percent by 2050.

TABLE 6.9 Gasoline Displacement for Cases 1-4 Compared to Reference Case

	Billion Gallons Gasoline Saved per Year (% Saved)		
Case	2020	2035	2050
Case 1 (Hydrogen Success)	1.0 (0.8%)	34 (24%)	109 (69%)
Case 2 (ICEV Efficiency)	2.2 (1.7%)	35 (25%)	64 (41%)
HFCVs + ICEV Efficiency	3.0 (2.2%)	55 (39%)	125 (80%)
Case 3 (Biofuels)	12 (9%)	28 (20%)	39 (25%)
Case 3 + Case 2 Biofuels + ICEV Efficiency	14 (11%)	64 (45%)	103 (66%)
Case 4: ICEV Efficiency (Case 2) + Biofuels (Case 3) + Hydrogen (Case 1)	15 (11%)	83 (59%)	157 (100%)

TABLE 6.10 Greenhouse Gas Emission Reductions for Cases 1-4 Compared to Reference Case

	Million Tonnes CO_2 Equivalent Avoided (% Avoided)		
Case	2020	2035	2050
Case 1 (Hydrogen Success)	10 (0.7%)	295 (19%)	1,026 (60%)
Case 2 (ICEV Efficiency)	24 (1.7%)	385 (25%)	700 (41%)
HFCVs + ICEV Efficiency	26 (1.8%)	475 (31%)	1,123 (66%)
Case 3 (Biofuels)	118 (8%)	281 (18%)	386 (23%)
Case 3 + Case 2 Biofuels + ICEV Efficiency	143 (10%)	666 (44%)	1,086 (64%)
Case 4: Hydrogen (Case 1) + ICEV Efficiency (Case 2) + Biofuels (Case 3)	130 (9%)	747 (49%)	1,505 (88%)

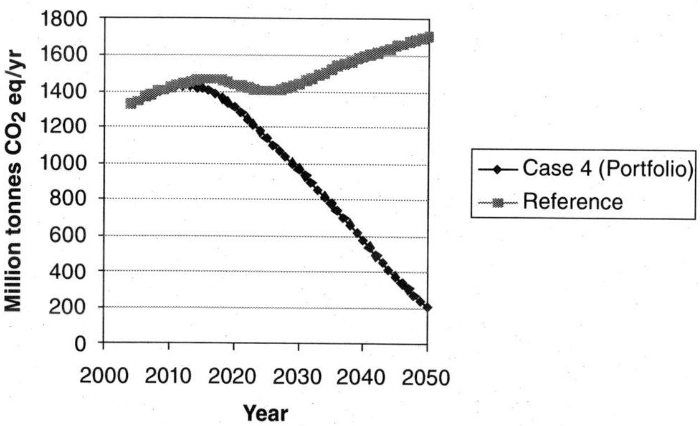

FIGURE 6.33 Greenhouse gas emissions for Case 4 (combination of HFCVs, efficiency, and biofuels).

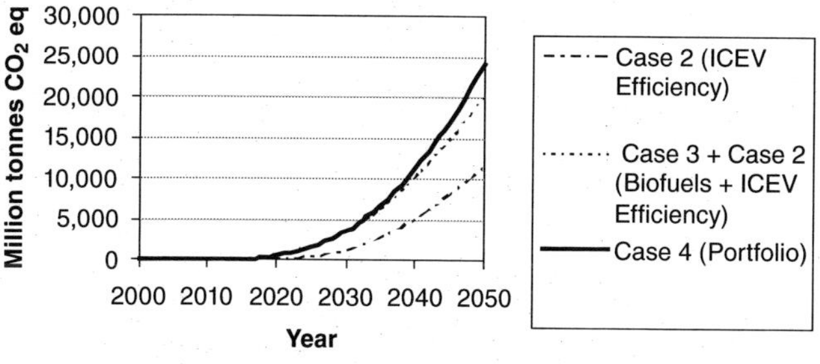

FIGURE 6.34 Cumulative reduction of greenhouse gas emissions for Case 2, Case 3 plus Case 2, and Case 4.

CONCLUSIONS

CONCLUSION: In the judgment of the committee, the maximum practicable number of HFCVs that could be on the road by 2020 is around 2 million. Subsequently, this number could grow rapidly to as many as 60 million by 2035 and more than 200 million by midcentury, but such rapid and widespread deployment will require continued technical success, cost reductions from volume production, and government policies to sustain the introduction of HFCVs into the market during the transition period needed for technical progress.

CONCLUSION: While it will take several decades for HFCVs to have major impact, under the maximum practicable scenario fuel cell vehicles would lead to significant reductions in oil consumption and also significant reductions in CO_2 emissions if national policies are enacted to restrict CO_2 emissions from central hydrogen production plants.

CONCLUSION: The unit costs of fuel cell vehicles and hydrogen in the Hydrogen Success scenario—the maximum practicable case—decline rapidly with increasing vehicle production, and by 2023 the cost premium for HFCVs relative to conventional gasoline vehicles is projected to be fully offset by the savings in fuel cost over the life of the vehicle relative to a reference case based on the EIA high-oil-price scenario. At that point, according to the committee's analysis, HFCVs become economically competitive in the marketplace.

Fully implementing the maximum practicable hydrogen case by 2050 would require construction of approximately 80,000 on-site distributed natural gas reforming units, 80 coal gasification plants of 500 MW (electrical equivalent) with CCS, 130 biomass gasification plants (each 100 MW equivalent) with associated biomass growth and collection farms, and roughly 80,000 miles of pipelines for hydrogen supply and CCS. The committee estimates that more than $400 billion would be required to fully build out hydrogen supply to fuel the HFCVs by 2050.

The committee's analysis indicates that at least two alternatives to HFCVs—advanced conventional vehicles and biofuels—have the potential to provide significant reductions in projected oil imports and CO_2 emissions. However, the rate of growth of benefits from each of these two measures slows after two or three decades, toward the end of the committee's analysis period, while the growth rate of projected benefits from fuel cell vehicles is still increasing. The deepest cuts in oil use and CO_2 emissions after about 2040 would be from hydrogen.

Over the next 20 years, the greatest impact on U.S. oil and CO_2 reductions would result from implementing existing evolutionary vehicles focused on vehicle design to deliver efficiency improvements.

The potential of biofuels under the committee's maximum practicable approach achieves a 23 percent reduction in CO_2 and gasoline use by 2050, but has only a small impact prior to 2035, compared to the reference case.

BIBLIOGRAPHY

Adrian, M. 2004. *Advanced Electronics and Control Technologies for Fuel-efficient, Low-emission Diesel Powertrains*. Doc. No. 2004-21-0083. Washington, D.C.: Greaney-Ricardo, Inc.

Argonne National Laboratory. 2007. The VISION Model. Available at http://www.transportation.anl.gov/software/VISION/index.html.

DOE (Department of Energy). 2005. Life-Cycle Analysis of Greenhouse Gas Emissions for Hydrogen Fuel Production in the United States from LNG and Coal. DOE/NETL-2006/1227, Table 5. National Energy Technology Laboratory, Pittsburgh, Pa.

DOE. 2007. H2A Hydrogen Analysis Model. Available at http://www.hydrogen.energy.gov/ h2a_analysis.htm.

EIA (Energy Information Administration). 2008. *Annual Energy Outlook with Projections to 2030*. Report DOE/EIA-0383. Washington, D.C.

Greene, D., P. Leiby, and D. Bowman. 2007. *Integrated Analysis of Market Transformation Scenarios with HyTrans*. Oak Ridge National Laboratory, Oak Ridge, Tenn.

Gronich, S. 2007. 2010-2025 Hydrogen Scenario Analysis. Presentation to the committee, February 20.

Kromer, M.A., and J.B. Heywood. 2007. *Electric Powertrains: Opportunities and Challenges in the U.S. Light-Duty Vehicle Fleet*. MIT Report LFEE 2007-02 RP, May. Cambridge, Mass.

Melendez, M. 2006. Geographically Based Hydrogen Demand & Infrastructure Analysis. Presented at the USDOE Hydrogen Transition Analysis Workshop, Washington, D.C., August 9-10.

Nicholas, M.A., and J.M. Ogden. 2007. Detailed Analysis of Urban Station Siting for California Hydrogen Highway Network. *Transportation Research Record* 1983:121-128.

Nicholas, M.A., S.L. Handy, and D. Sperling. 2004. Using Geographic Information Systems to Evaluate Siting and Networks of Hydrogen Stations. *Transportation Research Record* 1880:126-134.

NRC (National Research Council). 2008. *Review of the Research Program of the FreedomCAR and Fuel Partnership: Second Report*. Washington, D.C.: The National Academies Press.

Paster, M. 2006. Hydrogen Delivery Options and Issues. Presented at the USDOE Hydrogen Transition Analysis Workshop, Washington, D.C., August 9-10.

Perlack, R., et al. 2005. Biomass as Feedstock for a Bioenergy and Bioproducts Industry: The Technical Feasibility of a Billion-Ton Annual Supply. DOE-USDA, Washington, D.C.

Ruether, J., M. Ramezan, and E. Grol. 2005. Life-Cycle Analysis of Greenhouse Gas Emissions for Hydrogen Fuel Production in the United States from LNG and Coal. National Energy Technology Laboratory, Pittsburgh, Pa.

Sawyer, R. 2007. California's Regulations to Control Greenhouse Gas Emissions from Motor Vehicles. Presentation at the Hearing on Request for Waiver of Preemption under Clean Air Act Section 209(b) by Robert Sawyer, Chair, California Air Resources Board, Washington, D.C.

Singh, M., A. Vyas, and E. Steiner. 2003. VISION Model: Description of Model Used to Estimate the Impact of Highway Vehicle Technologies and Fuels on Energy Use and Carbon Emissions to 2050. Center for Transportation Research, Argonne National Laboratory, Argonne, Ill.

TIAX. 2007. *Full Fuel Cycle Assessment: Well to Tank Energy Input, Emissions and Water Impacts*. California Energy Commission Report, CEC-600-2007-002-D, February.

Yang, C., and J. Ogden. 2007a. Determining the Lowest-cost Hydrogen Delivery Mode. *International Journal of Hydrogen Energy* 32(2):268-286.

Yang, C., and J. Ogden. 2007b. U.S. Urban Hydrogen Infrastructure Costs Using the Steady State City Hydrogen Infrastructure System Model (SSCHISM). Presentation at the 2007 National Hydrogen Association Meeting, San Antonio, Texas, March 18-22. (A beta copy of the model is posted on Christopher Yang's website at University of California at Davis Institute of Transportation Studies, http://www.its.ucdavis.edu/people.)

7

A Budget Roadmap

As requested in the statement of task, this chapter estimates the required research, development, and demonstration (RD&D) funding, both by the federal government and by the private sector that would be required for a transition to hydrogen fuel cell vehicles (HCFVs). It also estimates the infrastructure and manufacturing costs to transition HFCVs into the market and considers whether a lack of skilled workers might inhibit development or deployment of the technologies required.

RESEARCH, DEVELOPMENT, AND DEMONSTRATION COSTS

Government RD&D Funding

The Bush administration and Congress committed $1.16 billion over 5 years starting in 2004 for RD&D of technologies for hydrogen fuel cell vehicles. This government effort, undertaken by the Department of Energy (DOE), had actual annual appropriations totaling $879 million through 2007, with $309 million requested for FY 2008. If appropriated, this would bring the federal program to the total committed amount. DOE program funding by focus area is shown in Table 7.1. State governments do not play a large role in funding hydrogen fuel cell vehicle technology, although several states are involved in infrastructure demonstrations.

Besides the obvious areas of focus, the DOE Nuclear Hydrogen Initiative is directed at using the high temperatures (about 900°C) of next-generation nuclear reactors in novel processes to free hydrogen from water. The DOE Science program is supporting basic research on novel membranes, catalysts at the nanoscale, novel materials for hydrogen storage, and new approaches for solar hydrogen production.

The statement of task direct the committee to make an estimate of future government RD&D funding for the transition from petroleum to hydrogen in a significant percentage of the vehicles sold by 2020. No hard data are available because DOE cannot publicize future spending plans beyond the current fiscal budget year. A rough estimate is made here out to 2020 when the number of fuel cell vehicles based on technologies developed in the Hydrogen Success case is projected to be nearly 2 million. Spending beyond 2020 would likely continue further improvements in the longer term—for example, in pursuit of hydrogen from renewable sources. Such spending is estimated in less detail through 2023 when fuel cell vehicles are projected to break-even competitively and become self-supporting. Funding estimates for DOE R&D directed at hydrogen light-duty vehicles are shown in Table 7.2 and total $4.1 billion through 2020 (constant 2005 dollars) plus an additional $0.9 billion for 2021-2023, bringing the overall total to $5.0 billion from 2008 to 2023. Other programs at DOE and in other agencies are not included in Table 7.2. These figures should be considered approximations only. It is quite likely that additional research and demonstration programs will be required.

Funding for distributed hydrogen production completes the work required to support initial hydrogen supplies during the transition, based on natural gas reforming and water electrolysis. Thus, the estimated budget includes demonstration projects (demos) of these technologies in service stations to foster technological and practical learning. The demos for each are 1,500 kg/d with capital and operating costs estimated based on DOE's H2A model as noted in Chapter 6 (DOE, 2007). These costs are $8 million for water electrolysis and $5 million for natural gas reforming, and they are included in the distributed production demo line of Table 7.2.

Centralized hydrogen production is the major source of hydrogen supply for larger fuel cell vehicle populations. Thus, the RD&D program comprises the development of hydrogen from coal gasification and biomass gasification, including demos to foster learning. Biomass estimates do not include crop research, which would be done anyway to support biofuels generally. Centralized production includes the hydrogen delivery infrastructure RD&D.

TABLE 7.1 Recent R&D Funding by the U.S. Department of Energy for Fuel Cells and Hydrogen Production (millions of constant 2005 dollars)

Program Area	2004	2005	2006	2007	Total
Hydrogen production and delivery	29.9	39.9	31.5	65.9	167.2
Fuel cells and hydrogen storage	52.7	68.2	58.7	89.6	269.2
Technology validation (learning demonstrations)	15.6	26.1	33.3	39.6	114.6
Safety, codes and education	8.2	5.8	5.1	15.8	34.9
System analysis	1.4	3.2	4.8	9.9	19.3
Nuclear Hydrogen Initiative	6.2	8.7	24.1	18.7	57.7
Science	0.0	29.2	32.5	50.0	111.7
Congressionally directed	42.0	40.2	42.5	0.0	124.7
Total	156.0	221.3	232.5	289.5	899.3

TABLE 7.2 Estimated Future Government Funding for RD&D (millions of constant 2005 dollars)

Program Area	2008	2009	2010	2011	2012	2013	2014	2015	2016	2017	2018	2019	2020	Total
Distributed H_2 production	12	15	8	8	3	0	0	0	0	0	0	0	0	46
Distributed H_2 production demonstrations	0	8	3	2	0	0	0	0	0	0	0	0	0	13
Centralized H_2 production	28	35	45	50	55	55	50	45	35	30	30	20	15	493
Centralized H_2 production demonstrations	0	0	0	15	15	15	15	50	35	20	25	20	0	210
Fuel cells and H_2 storage	112	115	115	115	115	110	110	110	110	110	110	110	110	1,452
Fuel cell demonstrations	30	40	40	50	40	30	30	20	20	20	15	10	10	355
Safety, codes, and education	21	21	25	25	25	25	15	10	10	5	5	5	5	197
Systems analysis	12	10	10	10	10	10	10	10	10	10	10	10	5	127
Science	60	60	60	60	60	60	60	60	60	60	60	60	60	780
Exploratory H_2 from renewables	34	35	35	30	30	30	30	30	30	30	30	30	30	404
Total, 2008-2020	309	339	341	365	353	335	320	335	310	285	285	265	235	4,077
Additional, 2021-2023														900
Total, 2008-2023														4,977

NOTE: Non-DOE programs are not included here, and some numbers are estimates that might change significantly with further information.

The learning demo for hydrogen from coal gasification is ancillary to a coal gasification plant under consideration by DOE and industry, with areas of responsibility and cost sharing to be determined. The plant, with carbon dioxide (CO_2) sequestration, would have a major goal of clean power production and would proceed independently whether or not it optionally produces pure hydrogen for fuel cell vehicles. Therefore, the cost to demonstrate hydrogen from coal gasification in the budget roadmap of this report is only the cost to demonstrate the incremental hydrogen purification required to meet fuel cell vehicle specifications. A very rough estimate of this cost is \$75 million. The biomass gasification demo is sized at 39,000 kg H_2/d, about one-quarter the size of a projected commercial plant. Capital and operating costs are estimated at \$135 million (constant 2005 dollars) based on the H2A model. The biomass and coal gasification demo costs are in the centralized production demo line of Table 7.2.

Major R&D efforts continue on the fuel cell to meet cost and durability targets. Hydrogen onboard storage work is directed at breakthroughs that would be much better than high-pressure hydrogen storage. Learning demos on fuel cells also are included in the budget roadmap.

DOE's Science program continues basic research to help meet hydrogen production and fuel cell vehicle goals and provides the basis for longer-term improvements. Work continues on improved fuel cell membrane materials for increased ion transport and membrane durability, catalyst design at the nanoscale for more efficient hydrogen production, and novel metal hydrides for hydrogen storage. Exploratory R&D on hydrogen from renewable energy sources includes the thermochemical splitting of water using high-temperature (900°C) energy from a new generation of nuclear reactors. This exploratory work also includes a photoelectrochemical process (in which photons absorb into an electrode and produce electrons for electrolysis) and a photobiochemical process (involving oxidative cleavage of water mediated by photosynthetic microorganisms). These approaches seek to liberate hydrogen from water with possible benefits compared to conventional water electrolysis, as well as longer-term benefits that would enhance the value of a transition to hydrogen-powered vehicles.

From 2021 to 2023 (the breakeven year for transition to HFCVs in the Hydrogen Success scenario), RD&D would follow on leads from the results of exploratory work on hydrogen from renewables and other areas of improvement that might surface as the transition proceeds. As for any major commercial venture, RD&D would be ongoing and continue beyond the 2023 breakeven point. It is likely that government support would still be involved, but longer-term funding levels are not projected here.

Private Sector RD&D Costs

The statement of task also requested an estimate of U.S. private RD&D funding for the transition to hydrogen fuel cell vehicles. This is very difficult to determine because such information is mostly proprietary. Private U.S. companies prominent in hydrogen fuel cell RD&D are GM, Ford, Chevron/Texaco, United Technologies, GE and nine venture capital companies listed on stock exchanges. These nine companies are Fuel Cell Energy, Quantum Fuel Systems, Plug Power, Medis Technology, Distributed Energy Systems, Hoku Scientific, Mechanical Technology, PolyFuel, and Protonex. These companies are noted in the Chapter 2 discussion of the role of entrepreneurial companies in the hydrogen transition. Together they had a combined invested capital of $2.5 billion at the year end of 2006. At least 32 other U.S. entrepreneurial companies also are active, albeit at a more modest level.

Some indications of the RD&D spending of private companies come from a 2005 worldwide hydrogen and fuel cell industry survey published in 2006. PricewaterhouseCoopers was the survey administrator that compiled and released the results in a pamphlet available on the Internet (PWC, 2007). This was a collaborative effort by the U.S. Fuel Cell Council, Hydrogen and Fuel Cells Canada, Fuel Cell Europe, and the Fuel Commercialization Conference of Japan. Unfortunately, the survey reported only total private and government spending, with no breakdown. However, those figures were used to estimate the private sector funding alone.

The survey reported total worldwide RD&D spending of $800 million in 2005, and the United States is reported at 40 percent of this amount, or $320 million. However, the response rate to the survey was 37 percent. If nonrespondents spent at the same rate as respondents, a "ballpark" estimate of U.S. RD&D spending in 2005 for private sector and government would be $860 million. U.S. government spending in 2005 was $220 million (see Table 7.1), thus leaving $640 million for the private sector, which represents roughly $700 million spent in 2005.

This rough estimate of $700 million per year was projected to continue from 2008 to 2023 for demonstrations, development of commercial-scale manufacturing, and addressing problems and opportunities that arise as the hydrogen transition proceeds. In total then, future U.S. private sector spending on RD&D for the hydrogen transition was estimated at roughly $9 billion through 2020 and $11 billion through the scenario breakeven year of 2023.

Conclusion

CONCLUSION: The committee estimates that total government-industry spending on RD&D needed to facilitate the transition to HFCVs is roughly $16 billion over the 16-year period from 2008 through 2023, of which about 30 percent (roughly $5 billion) would come from U.S. government sources. Government and private spending beyond 2023 also will likely be required to support longer-term needs, but such estimates were beyond the scope of this study.

INFRASTRUCTURE AND VEHICLE COSTS

As requested in the statement of task, this section estimates the total private and government funding that would be required to support a transition to hydrogen fuel cell vehicles, in addition to the RD&D funding estimated above. These budget estimates were prepared for activities that would be required for the Case 1 (Hydrogen Success) scenario described in Chapter 6.

Cost Elements

The major cost elements of a budget roadmap are summarized in Table 7.3. They include the capital requirements plus the annual operating and maintenance (O&M) costs for the two principal components of the system, namely (1) fuel cell vehicles and (2) hydrogen fuel supply for these vehicles. In the budget roadmap that follows, the committee assumed that

TABLE 7.3 Major Cost Elements in a Budget Roadmap

Cost Element	Capital Costs	O&M Costs
Vehicle production costs	Production facilities Testing facilities Base vehicle costs Fuel cell power train costs	Raw materials Labor costs (skilled, manual, supervisory) Facility operating costs (utilities, insurance, etc.) Facility maintenance costs Retailing costs
Hydrogen supply costs	Fueling station land and building costs H_2 supply technology (SMRs; coal gasification) Delivery system hardware (local or remote)	Feedstocks (natural gas, electricity, etc.) Labor costs Delivery costs Other operating costs Supply-related facility maintenance costs

NOTE: SMR = steam methane reformer.

TABLE 7.4 Projected Cumulative Infrastructure Requirements in 2020, 2035, and 2050 for the Hydrogen Success (Case 1) Scenario

	2020		2035		2050	
Infrastructure Items	Capital Cost (billion 2005 dollars)	Physical Units	Capital Cost (billion 2005 dollars)	Physical Units	Capital cost (billions 2005 dollars)	Physical Units
Hydrogen production plants:						
Natural gas on-site SMR stations	$2.6	2,110	$34	22,000	$122	79,000
Central coal plants with CO_2 capture and sequestration	0	0	$18.2	20	$80	79
Central biomass plants	0	0	$13.5	93	$19	131
Central natural gas SMR	0	0	0	0	0	0
Total production plant capital cost	$2.6		$66		$221	
Refueling stations	$2.6	2,110	$25	56,000	$74	180,000
Delivery system (dollars and pipeline miles)	0	0	$48	39,000	$120	80,000 H_2 pipelines; 5,000 CO_2 pipelines
Total cumulative capital costs	$2.6		$139		$415	

some of these costs would be borne by government programs during the transition period required for fuel cell vehicles to become competitive in the marketplace.

To estimate these costs, the results of the Case 1 scenario described in Chapter 6 were utilized. That scenario represents the committee's best estimate of the maximum practicable number of HFCVs that could be deployed in the United States by 2020 (and beyond). The committee first estimated the annual expenditures required to deploy the number of vehicles specified per year in that scenario. The vehicle production costs of Table 7.3 are captured in the estimated retail price of HFCVs developed in Chapter 6, and used here in the budget roadmap. The annual capital expenditures for infrastructure to supply hydrogen to those vehicles also were estimated, as well as the O&M costs of the infrastructure each year. The analysis of transition costs for this case ends in 2023, which is the breakeven year, after which HFCVs compete economically with conventional vehicles Hydrogen Success scenario (see Chapter 6).

Building the infrastructure needed to fuel HFCVs will be a substantial construction program in itself, aside from research and technology development. Table 7.4 shows the cumulative number of hydrogen production plants, refueling stations, and distribution pipelines, as well as their cumulative capital costs at several points in time. The committee's projected cumulative investment for hydrogen infrastructure

TABLE 7.5 Quantities Related to Infrastructure Estimates for the Hydrogen Success (Case 1) Scenario

Quantity	2020	2035	2050
Hydrogen demand (tonnes per day)	1,410	38,000	120,000
Hydrogen fuel cell vehicles served annually (% of total light-duty fleet)	1.8 million (0.7%)	61 million (18%)	219 million (60%)
Consumption of feedstocks for hydrogen production (exajoules per year):			
Natural gas	0.08	0.8	2.2
Coal	0	1.4	3.8
Biomass	0	1.6	2.2
CO_2 sequestered (million tonnes per year)	0	114	317

(Case 1) totals nearly $3 billion in 2020, and then climbs to $139 billion in 2035 and $415 billion in 2050. Although the estimated investment to build out the hydrogen infrastructure is clearly large, the committee's analysis, as explained in the pages that follow, assumes that the vast majority of the investment required will be made by industry because it is economically attractive to do so, and only a modest fraction will require government support. Table 7.5 reports several additional quantities related to these infrastructure projections, including numbers of vehicles served, amounts of hydrogen produced, energy feedstocks used, and CO_2 sequestered from central-station hydrogen production.

Total Annual Expenditures

Figure 7.1 shows the total annual expenditures involved for the purchase of fuel cell vehicles and the production of hydrogen. The deployment of HFCVs starts with 1,000 vehicles in 2012, increasing to commercial levels of 50,000 per year in 2015, 750,000 per year in 2020, and 1.5 million per year in 2023—the breakeven year, after which the HFCV market is self-sustaining in the Hydrogen Success scenario. During this period the unit price per vehicle falls from slightly more than $200,000 in 2012 to just over $27,000 in 2023, as indicated in Chapter 6. The resulting total annual expenditures for the 5.5 million vehicles deployed by 2023 are shown in Figure 7.1, along with the average per-vehicle price ($30,000), the additional capital expenditures for hydrogen supply infrastructure, and the annual O&M costs of hydrogen production (mainly for natural gas feedstock).

The total annual expenditures for vehicles and hydrogen supply in Figure 7.1 increase from about $300 million in 2012 to $46 billion in 2023, with the cumulative expenditure over the transition period reaching $184 billion in 2023. Most of that amount (91 percent) is for the purchase of fuel cell vehicles. The remaining 9 percent ($16 billion) is for hydrogen supply, divided about equally between the capital costs of hydrogen infrastructure and the O&M costs for hydrogen production. The annual vehicle costs shown here reflect all of the capital and O&M cost elements shown in Table 7.3. Some of those elements, however (such as the capital investment for fuel cell production facilities), are reflected only implicitly as part of the per-vehicle price estimates used here. A more detailed budget roadmap showing the breakdown of estimated annual expenditures for all of the individual cost elements in Table 7.3 (e.g., production facilities, equipment, and raw materials over the transition period) is well beyond the scope of the present analysis.

Note that since the budget roadmap of Figure 7.1 shows

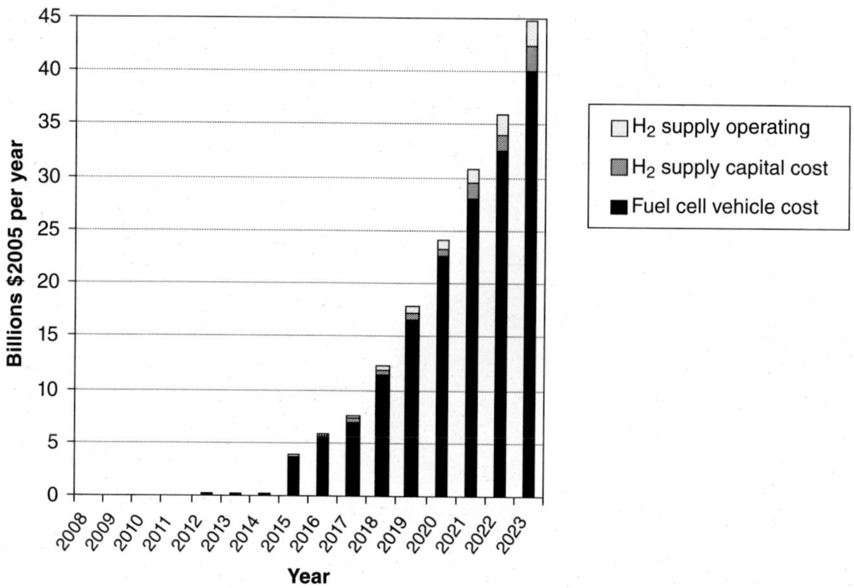

FIGURE 7.1 Total annual expenditures for vehicles and hydrogen supply for transition to the breakeven year for the Hydrogen Success case, excluding RD&D costs. The cumulative cost, shared by government and industry, totals $184 billion, of which 91 percent is the cost of fuel cell vehicles and 9 percent is the cost of hydrogen supply (about half for infrastructure costs and half for additional operating costs, mainly natural gas feedstock).

only the total private plus government expenditures needed to implement the Case 1 scenario, it does not reflect the $17 billion net savings in consumer expenditures for fuel from 2012 to 2023 (discussed in Chapter 6) as hydrogen-fueled vehicles become more competitive with gasoline vehicles. Nor does it reflect the roughly $5 billion loss of federal and state government tax revenues from gasoline sales displaced by hydrogen (which is assumed to be free of taxes in this analysis).

Government Versus Industry Funding

The question of how the total annual costs shown in Figure 7.1 should be shared between the federal government and private industry has no simple or single answer. Conceivably, the government could bear all of the $184 billion in vehicle and hydrogen supply costs through 2023 to accelerate the deployment of fuel cell vehicles. This situation might apply if the technical and market readiness of HFCVs was perceived by industry as still too risky to warrant private investments of the magnitude required over this time frame. Thus, government would have to bear all of the costs and risks as the de facto customer for all HFCVs. The committee believes that such a scenario is unrealistic since major auto companies would not likely be willing to commit facilities and personnel, or risk their reputation and current development plans, on a venture they perceive as too risky, even if government offered to pay the bill. As discussed below, however, the government might buy a substantial fraction of new HFCVs in the early years of the transition for use in its own fleet.

In the committee's judgment, a realistic estimate of the government share of total costs to facilitate the maximum practicable transition to HFCVs (based on the Hydrogen Success case) would be the incremental cost of purchasing fuel cell vehicles, plus about half the total cost of building and operating the infrastructure needed to supply hydrogen during the transition period (the remaining half is assumed to be provided by the private sector). In practice, it is desirable that industry share the costs of both constructing and operating the hydrogen supply system. However, since the cumulative costs for infrastructure construction and operation are approximately equal ($8 billion each over the transition period), the committee assumed for simplicity that all capital costs are borne by government and all operating costs by the private sector. These incremental costs are shown in Figure 7.2. In this case, the cumulative government expenditure for vehicles totals $40 billion over the transition period, as noted in Chapter 6, while hydrogen supply costs add another $8 billion, bringing the total to $48 billion. This amounts to 26 percent of the $184 billion in total expenditures for vehicles and hydrogen supply over the transition period in Case 1.

To the extent that the Case 1 deployment schedule for HFCVs succeeds in meeting or exceeding the technical and cost targets assumed in this analysis, the government's share of total costs could be reduced further relative to the budget roadmap of Figure 7.2. Some consumers also may be willing to pay a premium for this new type of vehicle. On the other hand, to the extent that early program goals are not fully achieved, or industry is reluctant to commit to the deployment schedule assumed in this analysis, greater government funding would be required to sustain the Case 1 scenario. For example, if government bore the full vehicle cost, rather than the incremental cost, during the first 5 years of production, it would add about $4 billion to the total cost for approximately 150,000 HFCVs. *For reference, this is about half the number of new vehicles currently purchased*

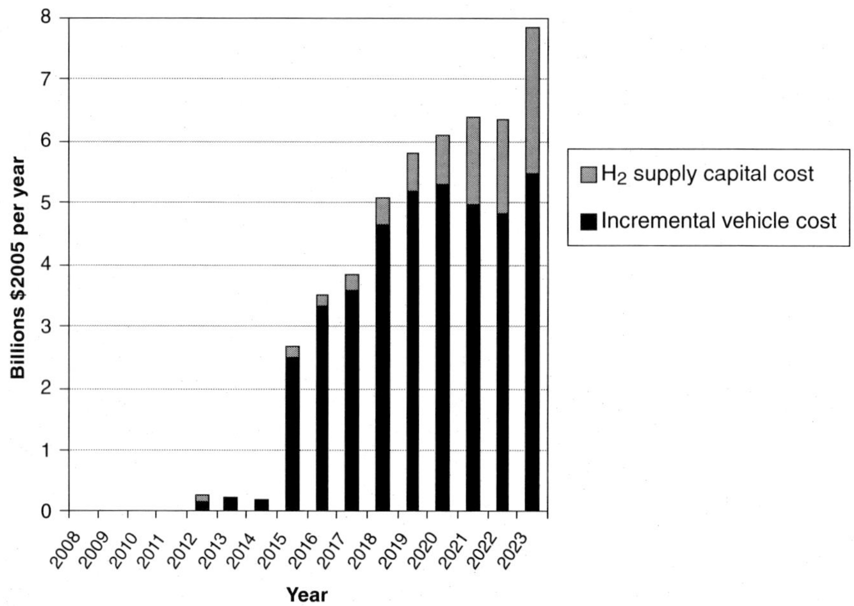

FIGURE 7.2 Annual government expenditures through the transition to 2023. Estimated expenditures are based only on the incremental costs of fuel cell vehicles over conventional vehicles, plus the capital cost for hydrogen infrastructure, for the Hydrogen Success scenario (excluding RD&D costs). The cumulative cost is $48 billion, of which 83 percent is the cost of vehicles and 16 percent is the cost of hydrogen infrastructure. Government RD&D costs over this period total an additional $5 billion.

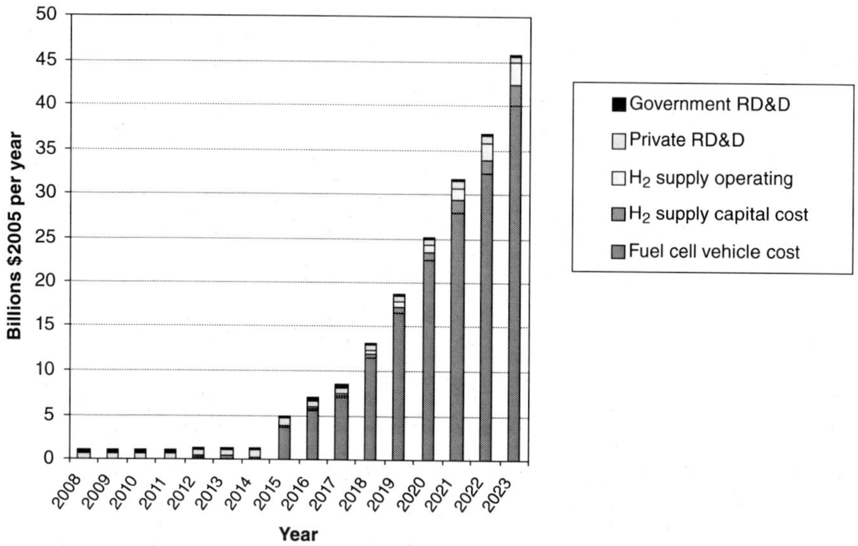

FIGURE 7.3 Total annual costs of transition to the breakeven year for the Case 1 scenario, including RD&D costs plus total vehicle and hydrogen supply costs.

by the federal government over a 5-year period.[1] *Allowing for some expenditures of this nature,* ***the committee estimates the government share of total vehicle plus hydrogen costs to be approximately $50 billion (an average of $9,500 per vehicle) during the transition period.***

The committee's analysis assumed (for simplicity) that all costs shown in Figures 7.1 and 7.2 are borne by U.S. companies and government. To the extent that participation by Japanese and other foreign manufacturers accelerates the introduction of HFCVs, and subsidizes the costs of a transition to fuel cell vehicles, the total U.S. costs shown here would be further reduced. For example, if early HFCV markets outside the United States were half as large as the assumed U.S. markets, the time for transition would be accelerated by 1-2 years, and the cumulative cost difference between HFCVs and gasoline vehicles would be reduced by $5 billion to 10 billion (from the $40 billion estimated here) assuming shared learning. Although the committee did not attempt to estimate the potential role of non-U.S. investments in HFCV technologies, it is aware that major efforts outside the United States are currently under way and could have a significant influence on the development and cost of a transition to HFCVs in this country.

OVERALL BUDGET ROADMAP

Figures 7.3 and 7.4 combine estimates of government and private sector RD&D costs with the estimated costs of vehicle deployment in Figures 7.1 and 7.2, respectively. The overall cost for the transition period (2008 to 2023, inclusive) then totals approximately $200 billion, shared between industry and government. ***The government portion of the total transition cost, including RD&D, is estimated to be roughly $55 billion (an average of $10,000 per vehicle),*** as summarized in the last line of Table 7.6 (which shows estimated transitions costs on a cumulative and average per-vehicle basis). As discussed above, these estimates are based on the committee's Hydrogen Success scenario defining the maximum practicable number of HFCVs that could be on U.S. roads by 2020. This overall cost range translates to an average of roughly $3 billion per year over 16 years (2008-2023). To put these amounts in perspective, the U.S. government subsidy for ethanol fuel in 2006 was approximately $2.5 billion and, if extended at the current rate, could grow to $15 billion per year in 2020 as a result of the recent (December 2007) energy act.[2]

Note, too, that while the budget roadmaps presented here apply only to the transition period through 2023, the successful introduction of fuel cell vehicles would involve substantial additional expenditures—primarily by the private sector—for infrastructure, energy resources, and other requirements of a full-scale HFCV-based transportation system. For example, as seen in Table 7.4, the committee estimated that over more than $400 billion would be required by 2050 to fully build out the hydrogen supply system to fuel the HFCVs. However, the committee believes that follow-

[1]The overall federal fleet is about 650,000 vehicles, with acquisitions of about 65,000 per year. While many of the newly acquired vehicles would not be appropriate for hydrogen or would not be in an area where hydrogen is available, the federal fleet could by itself account for a significant fraction of early HFCVs (GSA, 2007).

[2]The Volumetric Ethanol Excise Tax Credit (VEETC) of 51 cents per gallon is provided to all ethanol blended with gasoline, which was about 5 billion gallons in 2006, according to DOE data. Although the VEETC is set to expire after 2010, Congress is debating various ways of extending it, as it has since the credit was first created in 1978. The Energy Independence and Security Act of 2007 established a renewable fuel standard that would reach 30 billion gallons by 2020, most of which is likely to be ethanol. A 51 cents per gallon credit applied to that amount would represent a subsidy in excess of $15 billion per year.

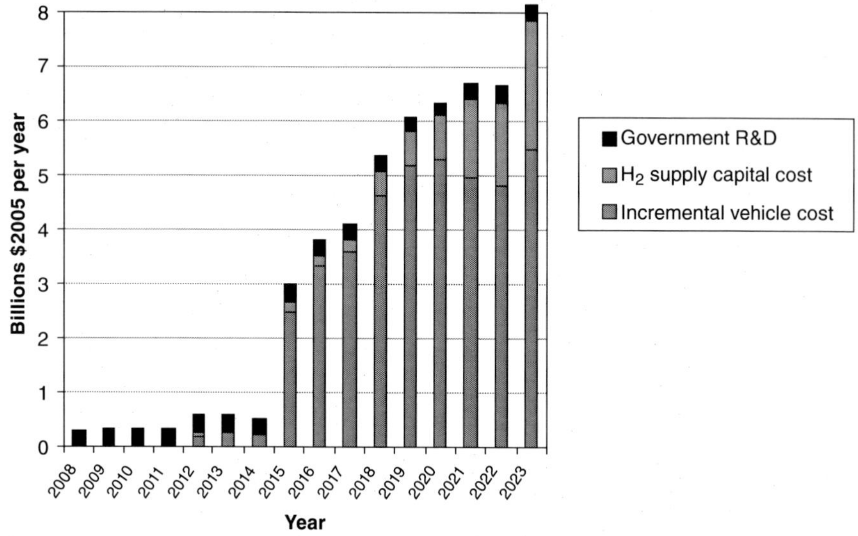

FIGURE 7.4 Total annual costs of RD&D plus incremental costs of HFCVs over conventional vehicles up to the breakeven year for the Case 1 scenario.

TABLE 7.6 Summary of Cumulative Budget Roadmap Costs for Transition to Hydrogen Fuel Cell Vehicles (maximum practicable number of vehicles by 2020)

Cost Elements	Total Cumulative Cost, 2008-2023	Average Cost per HFCV on Road 2008-2023[a]
"Base vehicle" cost of conventional vehicles	$128 billion	$23,000
Average incremental fuel cell vehicle cost relative to conventional gasoline vehicles	$40 billion	$7,000[b]
Total purchase cost of fuel cell vehicles	$168 billion	$30,000[c]
Infrastructure capital cost for hydrogen supply	$8 billion	$1,500
Total operating cost for hydrogen supply	$8 billion	$1,500
Total cost of hydrogen supply	$16 billion	$3,000
Total cost for vehicles and hydrogen fuel supply	$184 billion	$33,000
Estimated government share of total vehicle and hydrogen fuel supply cost	$50 billion	$8,500
Government RD&D funding	$5 billion	$1,000
Private RD&D funding	$11 billion	$2,000
Total funding for government and private RD&D	$16 billion	$3,000
Total cost for vehicles, hydrogen, and all RD&D	$200 billion[d]	$36,000
Estimated government share of total cost for vehicles, hydrogen, and RD&D	$55 billion	$9,500

[a]Rounded estimates based on 5.54 million HFCVs on the road in 2023.
[b]The final (learned-out) incremental cost per vehicle in 2023 is $3,600.
[c]The final (learned-out) cost per vehicle in 2023 is $27,000.
[d]Includes $128 billion "base vehicle" cost of conventional vehicles that would have been purchased instead of HFCVs.

NOTE: All costs in constant 2005 U.S. dollars.

ing a successful transition, there would be sufficient market incentives for industry to invest the needed capital without government support. Additional details of some of the longer-term resource needs to support the Hydrogen Success scenario are found in Chapter 6.

Other factors besides technical progress and funding requirements could also affect the viability of achieving the Hydrogen Success scenario. While a comprehensive assessment of potential barriers to a transition to fuel cell vehicles was beyond the scope of this study, one of those factors—the availability of the requisite skilled workforce—was examined by the committee, as discussed below.

SKILLS AVAILABILITY

The question looms large as to whether there will be sufficient professional and skilled labor resources available to achieve the growth in fuel cell vehicles reflected in the scenarios developed by the committee. Numerous recent reports have highlighted concerns about the availability of critical skills and trained personnel resources, particularly in the energy arena. For example, a recent National Petroleum Council report declared that a demographic cliff is looming in all areas of energy industry employment (NPC, 2007). The U.S. Department of Labor recently released a report on the "graying" of workers in the nuclear industry that indicated around one-third of the workforce in that segment of the energy business will be eligible to retire in the next 5 years (EBiz, 2007). The committee also is aware that DOE has been conducting an employment study related to the hydrogen and fuel cell industry, but that study has not yet been released and was not available to the committee during the committee's information-gathering efforts.

To address the question of skills availability in the hydrogen and fuel cell industry, the committee constructed a simple flow diagram showing the elements of this emerging industry segment (Figure 7.5). Key industrial players in each element of the business model shown in Figure 7.5 were identified, and more than 20 "not-for-attribution" interviews were conducted with executives representing those companies (as well as with several academics and nongovernmental organization officials with expertise in the field). The companies ranged from early-stage entrepreneurial businesses to international giants with global reach. Uniformly, the feedback received in these interviews was that there was little or no concern about attracting the skills needed to achieve the growth trajectories envisioned from the initial commercial introduction of hydrogen vehicles until 2025—even in the most aggressive of the committee's scenarios.

It is possible that beyond 2025, when the committee's projections envision the construction of large facilities for central generation of hydrogen, as well as pipeline delivery systems, labor force constraints may become more critical, but the committee is reluctant to speculate on that possibility at this time. It was also recognized by the committee that new skills and knowledge will have to be acquired by public sector agencies dealing with codes, standards, and safety matters related to the hydrogen transition. Planning and effort will be required well in advance to ensure that the skills and knowledge needed to permit, install, and operate hydrogen systems will be available when necessary.

In summary, the committee found no evidence from the information gathered in its brief field research effort that there is any need to be concerned about the availability of critical skills to achieve a rapid ramp-up in the introduction of fuel cell vehicles and related fueling infrastructure, at least

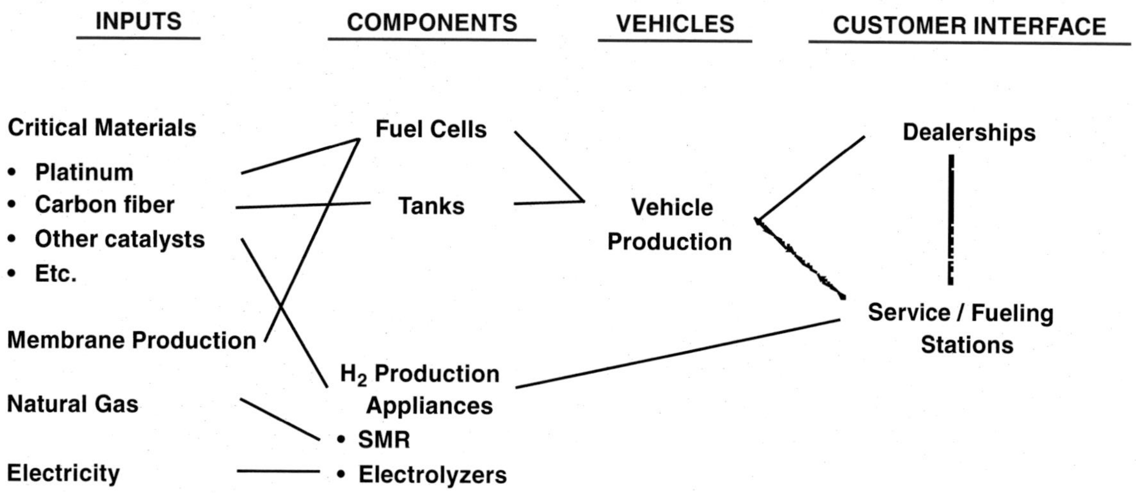

FIGURE 7.5 Diagram of the early structure of the hydrogen and fuel cell industries, identifying areas where skilled people will be needed. NOTE: SMR = steam methane reformers.

in the period until 2025. Rather, the consensus among the sample of executives interviewed by the committee indicated that attracting the skills needed to achieve any foreseeable rate of growth in the hydrogen and fuel cell industry will not be a problem, particularly if the government sends a clear signal that its support for growing the industry will be substantial and enduring.

It is also interesting to speculate on why this conclusion was found for hydrogen and fuel cell vehicles when other segments of the energy industry are expressing concern about skills availability. Three factors were suggested by the interviews conducted:

1. First, the scale of the hydrogen and fuel cell industry in the early years, even at the most rapid pace of growth envisioned by the committee, is still relatively small compared to the world's overall energy supply infrastructure.
2. Much of the leading work being done in the hydrogen and fuel cell industry is coming from special task groups set up by divisions of major corporations (frequently referred to as "skunk works") or from entrepreneurial companies. These environments typically draw the best and the brightest scientists, engineers, and skilled technicians for reasons such as challenge, excitement, and opportunity for substantial economic gain.
3. A refrain frequently heard in interviews and contacts by committee members with professionals in the hydrogen and fuel cell field is that working in this field offers various intangible benefits to employees—a feeling of doing something good, helping to address a critical global problem, creating a new industry, and so on. These intangible benefits appear to be attracting many of "the best and the brightest" to the field.

CONCLUSION

CONCLUSION: The estimated government cost to support a transition to hydrogen fuel cell vehicles is roughly \$50 billion over a 16-year period from 2008 to 2023, primarily for the production of fuel cell vehicles (\$40 billion of incremental cost) and, to a lesser extent, for the initial deployment of hydrogen supply infrastructure (about \$10 billion) and R&D (about \$5 billion). No shortages are foreseen in the critical workforce skills needed to accomplish the transition. However, further study is necessary to assess the longer-term costs, institutional issues, workforce issues, and impacts of undertaking the major hydrogen infrastructure development required to support widespread use of HFCVs.

REFERENCES

DOE (U.S. Department of Energy). 2007. Hydrogen Fuel Cells and Infrastructure Technologies Program. Available at www.eere.energy.gov/hydrogenandfuelcells/analysis/model.html.

EBiz. 2007. Nuclear Jobs. energyBizinsider, September 14, 2007. Available at http://www.energycentral.com/site/newsletters/ebi.cfm?id=383.

GSA (General Services Administration). 2007. Available at http://www.gas.gov/graphics/ogp/FFR2007_508.pdf. Accessed April 2008.

NPC (National Petroleum Council). 2007. *Facing the Hard Truths About Energy—A Comprehensive View to 2030 of Global Oil and Natural Gas*. Washington, D.C.

PWC (PricewaterhouseCoopers). 2007. 2006 Worldwide Fuel Cell Industry Survey. Available at http://www.pwc.com/servlet/pwcPrintPreview?LNLoc=/extweb/pwcpublications.nsf/docid/6F870010939851E0852570CA00179123.

8

Actions to Promote Hydrogen Vehicles

As discussed in Chapters 1 and 2, the key motivations for this report, and for the interest in hydrogen-powered vehicles, are that increasing imports of petroleum leave the nation vulnerable to disruptions and high costs in the transportation sector, and that emissions of carbon dioxide (CO_2) from vehicles may have to be curtailed to avoid serious climate change. This report assesses the resource needs to accelerate a transition to hydrogen fuel cell vehicles (HFCVs) by 2020 and the resulting impacts on reductions in U.S. oil use and CO_2 emissions. However, such a transition to HFCVs is unlikely to happen by itself. As discussed in Chapters 6 and 7, too many changes have to occur in manufacturing and fuel supply, with prospects for profit too remote for private enterprise to make investments at the rate necessary to meet national objectives. Therefore a significant federal role will be essential. As requested in the statement of task, this chapter examines policy needs and options required to deploy the "maximum practicable" number of HFCVs discussed in earlier chapters.

GENERAL POLICY APPROACHES

Policies can be targeted directly at HFCVs (push) or aimed more broadly at creating an environment that favors them (pull). The committee concluded that although broad measures that allow flexibility for the market to find the best options are usually the best approach, this will not be adequate to implement HFCVs. General measures to reduce oil use and greenhouse gases, such as energy taxes, CO_2 taxes, and greenhouse gas cap-and-trade systems, have a vital role to play in encouraging a broad array of options for reducing emissions and oil use across the economy. However, they will do little on their own to encourage commercialization of transformative technologies, such as hydrogen, for the foreseeable future.

For example, current bills in Congress would impose an effective price of around \$5 to \$20 per ton of CO_2 emissions, which—given the carbon content of gasoline—is equivalent to about 5-18 cents per gallon of gasoline. This fuel price increase would raise the (discounted) life-cycle costs of operating new gasoline vehicles by roughly \$200-\$700, which is a small fraction of the difference in life-cycle costs between hydrogen and gasoline vehicles in the early years of the alternative hydrogen pathway scenarios in Chapter 6. Even if a long-term commitment to an aggressive greenhouse gas cap-and-trade system, radically higher fuel taxes, or more stringent fuel economy standards were credible, these alone may not be sufficient to spur a transition to HFCVs. For example, auto companies may be reluctant to incur large upfront costs in alternative vehicle development if some of the benefit spills over to rival firms that might be able to imitate new technologies or use them to further their own research and development (R&D) programs.

While the committee concluded that a portfolio approach to R&D is necessary to ensure that a sufficient number of options are available to reduce U.S. oil consumption and carbon dioxide emissions, policies to implement specific options will have to be tailored to the technology or outcome desired. For example, improved fuel economy of conventional vehicles may be achieved by raising corporate average fuel economy (CAFE) standards, as recently enacted by Congress, and greater use of biofuels can be encouraged with economic incentives such as those currently in place for ethanol. The committee believes, however, that the current barriers to deploying large numbers of fuel cell vehicles are too great for existing policies to produce rapid growth in HFCV deployment. Thus, if the U.S. government wants to have 2 million HFCVs operating by 2020, it must employ targeted policies designed to push HFCVs into the marketplace. Examples of such policy options are discussed below.

POLICIES SPECIFIC TO HYDROGEN FUEL CELL VEHICLES

If policy makers decide that the technology is sufficiently developed to warrant promoting long-run penetration of hydrogen vehicles into the in-use vehicle fleet, this committee believes that technology-push incentives that are carefully targeted, substantial, durable, coordinated, and progressively phased out over time will be required.[1]

- Targeted incentives, such as federal tax credits or subsidies for hydrogen vehicle purchase, or minimum sales share quotas imposed on manufacturers, would be needed to kick-start the market for hydrogen vehicles, given that these vehicles cannot become commercially viable until extensive learning-by-doing has occurred through sequential vehicle development, and the vehicles go into the mass production needed to achieve scale economies.
- Substantial incentives would be required, given that life-cycle costs for hydrogen vehicles are currently so much higher than those for comparable gasoline vehicles. Furthermore, without large incentives, consumers may be reluctant to switch to a new, untested vehicle they are not familiar with, while producers competing in global markets may be reluctant to undertake major and risky investments in transitional technologies.
- Durable incentives, lasting 15-20 years or more, would be critical for altering private sector expectations about the long-run payoffs to investments with high up-front costs.
- Coordination of incentives would also be important; for example, even if there is a substantial subsidy for hydrogen vehicle sales, auto manufacturers may still not invest in the technology if incentives for required infrastructure investment, or continued basic R&D, are perceived as inadequate. This is a chicken-and-egg problem: vehicle manufacturers will not produce the vehicles without knowing that the hydrogen supply will be there, while hydrogen producers will not supply the fuel without knowing that the demand for it will be there.
- Finally, any subsidies should be progressively phased out over time as long-term penetration targets are approached; this limits funding requirements from the government and encourages firms to act more quickly to obtain larger subsidies offered in the earlier years of the program.

The committee also emphasizes that a decision to aggressively push hydrogen must be based on an assessment of the relative risks and benefits of HFCVs, as well as competing technologies. On the one hand, the sooner a hydrogen program is initiated, the greater is the likelihood of achieving large reductions in gasoline consumption and the greater the likelihood that rapidly industrializing countries might transition to hydrogen vehicles before getting locked into conventional gasoline.[2] On the other hand, there is the downside risk of pushing HFCVs (or any other specific technologies) before they are really ready or if they turn out not to be the best option, which could be extremely expensive and disruptive. The committee concluded that HFCV technology, while promising, is still in the R&D phase and is not yet ready for a decision to initiate commercialization. That decision may come in a few years, and, as shown in Figure 7.1, expenditures would start to increase rapidly in 2015. Thereafter, the technology must be reassessed every few years during the transition, and policies readjusted, if progress on reducing costs and improving performance differs from expectations. In keeping with the requirement that policies be durable, however, any potential readjustment must not appear to threaten the long-term investments that must be made by industry. Policies must be carefully crafted to be both believable and realistic.

PROS AND CONS OF SUBSIDIES AND QUOTAS

Although government support for basic and applied R&D and fuel distribution infrastructure is necessary, the heart of any program to ultimately promote substantial hydrogen vehicle penetration must be the incentives for auto manufacturers to develop and mass-produce hydrogen vehicles. The two most direct ways to achieve this include a pricing-based strategy, where the market price of hydrogen vehicles is initially heavily subsidized by the government, and a quantity-based approach involving a progressively more stringent sales share requirement for hydrogen vehicles imposed on auto manufacturers. This section discusses the pros and cons of the price- and quantity-based approaches. A final section comments very briefly on broader, technology-neutral approaches to reducing oil use and CO_2 emissions.

Price-based Approach (Subsidies)

A price-based approach involves adoption of a schedule of government subsidies, such as the one sketched in Figure 8.1 (which corresponds to the Hydrogen Success scenario in Chapter 6) based on a specified difference in price between fuel cell vehicles and conventional vehicles. From a policy maker's perspective, a major drawback of the price-based approach is that the future vehicle penetration rates under a fixed, declining schedule for hydrogen vehicle subsidies are very uncertain, because they will vary with future market developments. For example, vehicle penetration might be rapid and substantial if hydrogen vehicle technology evolves quickly, but if not, and/or if consumers and firms are especially reluctant to embrace hydrogen vehicles, penetration rates might be minimal. Put another way, we simply cannot know in advance what subsidy schedule is required to meet a particular future target for hydrogen vehicle penetration.

[1]Moreover, if any transition to hydrogen vehicles is to greatly reduce CO_2 emissions over the longer haul, plants supplying the hydrogen must incorporate low-carbon technologies such as renewable hydrogen or fossil hydrogen with carbon capture and storage technologies.

[2]There is enormous potential for future growth in vehicle ownership in countries such as China and India. Vehicle ownership rates are currently less than 10 per thousand in both of these countries compared to more than 700 per thousand in the United States.

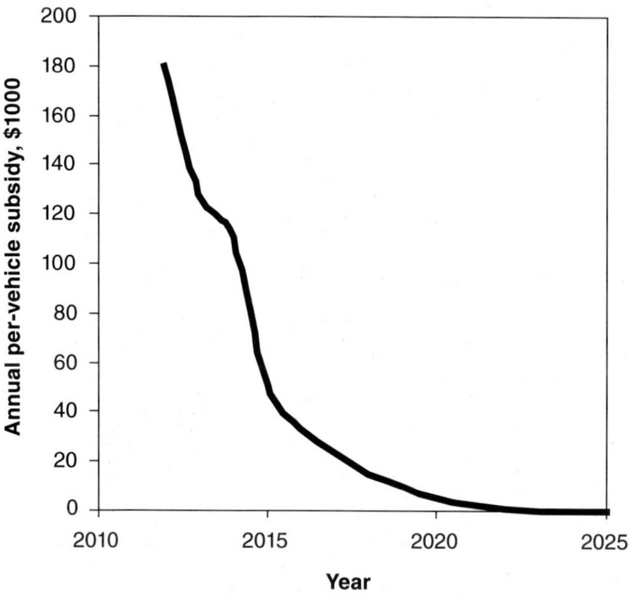

FIGURE 8.1 Illustrative example of a price-based policy approach, indicating the per-vehicle subsidy from government for each fuel cell vehicle sold in a particular year for the Hydrogen Success (Case 1) scenario.

One way to reduce the uncertainty over future penetration rates is to use a discretionary policy whereby subsidy levels or tax credits are adjusted upward or downward, according to whether future penetration rates turn out to be below or above target levels. In other words, the policy is adjusted to try and keep the net-of-subsidy life-cycle costs of hydrogen vehicles below (perhaps well below) those for comparable gasoline vehicles in any given year. However, this can be problematic in several respects. In particular, it may make more sense to adjust the penetration targets—for example, make them less ambitious, or possibly even abandon them altogether, if hydrogen technology evolves at a rate slower than that for others, such as biofuels or electric vehicles. It also creates an uncertain environment for investment decisions if firms do not know when, or by how much, future subsidies might be revised, and it may even have perverse effects if firms anticipate that their progress on vehicle development may cause future subsidies to be cut. A possible compromise might be to allow some very limited flexibility, for example, a midterm review of progress with a once-and-for-all correction in the subsidy schedule, based on criteria clearly specified in the initial legislation.

Quantity-based Approach (Quotas)

In contrast to the price-based approach, a quantity-based approach would impose a rigid sales share quota, such as illustrated in Figure 8.2 (which also corresponds to the Hydrogen Success scenario in Chapter 6). This approach might be more appealing to policy makers because it achieves a given hydrogen penetration target with far more certainty. Here, manufacturers must sell hydrogen vehicles through their own vehicle pricing strategies, regardless of market conditions and competition from other types of vehicles. The quota might be accompanied by a subsidy to manufacturers to assist them in getting through the very expensive transition, but the driving force would still be the quota. As suggested by the two figures, either approach can achieve the same results. The main drawback of this approach is that there is no limit on the costs of the policy, which could be especially burdensome if technological advance is slower than for competitor vehicles and hydrogen vehicles remain relatively costly to produce. In contrast, a (fixed) subsidy under the price-based approach is more flexible because it provides a natural mechanism for capping program costs, as manufacturers are free to scale back any plans for new vehicle production if future market conditions do not favor hydrogen vehicles. Imposing a uniform, minimum sales share quota across different manufacturers can also be problematic if it is relatively easy for some firms to meet the quota (for example, firms that are further ahead on the learning curve) and relatively costly for others. Again, this is not an issue under the subsidy policy since low-cost hydrogen vehicle producers will take more advantage of the subsidy by selling more vehicles, while high-cost producers will sell fewer vehicles and forgo the subsidy. However, the problem

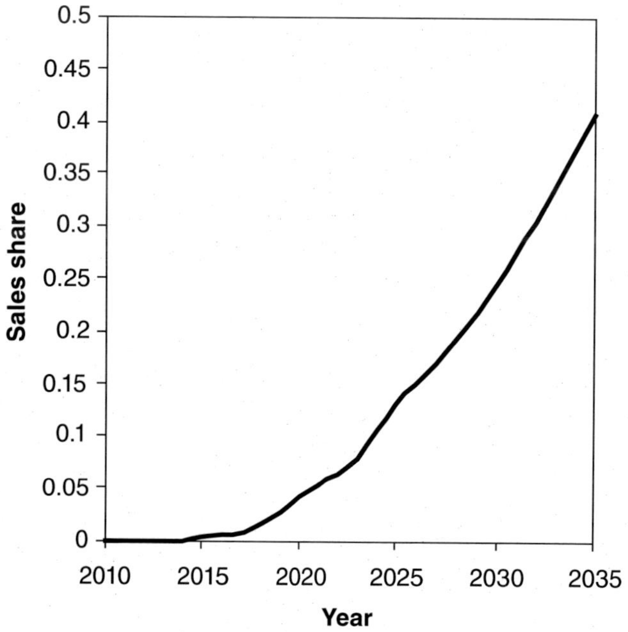

FIGURE 8.2 Illustrative example of a quantity-based policy approach, indicating the required fraction (quota) of all new vehicles sold in a particular year that must be fuel cell vehicles for the Hydrogen Success (Case 1) scenario.

under the quantity-based approach could be addressed by allowing firms to trade credits for hydrogen vehicle sales among themselves so high-cost manufacturers can opt to be below the standard by purchasing credits from other firms that exceed the standard.

The quantity-based (quota) approach may also face far more resistance from auto manufacturers, because they (or their customers) must bear the losses from selling hydrogen vehicles while their life-cycle costs are still above those for comparable gasoline vehicles; in contrast, under the subsidy approach these losses are ultimately borne by the general taxpayer.

Combined Approaches

The differences between price- and quantity-based approaches illustrates a fundamental tension between meeting a future penetration target for hydrogen vehicles with more certainty, and keeping down the possible costs of the program. A further option is to combine elements of the two approaches in a hybrid policy. For example, a relatively modest quota on hydrogen vehicles could be combined with subsidy inducements to go beyond the quota, which producers will take advantage of if future market developments favor hydrogen vehicles. Alternatively, a fairly stringent sales share quota could be mandated, but combined with a "safety valve" that allows manufacturers to pay a penalty in lieu of meeting the standard, which they will take advantage of if future market developments are more favorable to other technologies than to hydrogen.

Phased Hydrogen-specific Approach

One way to reconcile the contradiction between wanting to move ahead rapidly but lacking sufficient confidence that hydrogen is the "right choice" would be a phased approach; the decision to scale up the policy to a national level would be contingent on the success of "lighthouse" pilots, as discussed in Chapter 6. In this regard, a government procurement approach, where state and municipal government vehicle fleets transition to HFCVs, might play a useful role in helping with early development of the market. Such an approach would provide a way to avoid rapidly escalating costs if the technology did not advance as expected, but it would also introduce uncertainty in private sector planning.

BROAD POLICIES TO REDUCE OIL USE AND GREENHOUSE GAS EMISSIONS

Whether or not policy makers choose to push HFCVs aggressively, there are a variety of broader (technology-neutral) options to reduce greenhouse gas emissions and oil use that have been implemented, or are under consideration, at both federal and state levels in the United States, and in other countries. These broader polices do not attempt to pick a technology, which, as we know from past experience, might significantly distort the market. Some of these policies are sector specific, whereas others are economy-wide; some are price based and others are quantity based; some are based on performance standards.

For example, fuel economy or vehicle greenhouse gas standards that increase steadily in stringency over time ensure that emerging efficiency technologies are applied to fuel economy improvements instead of acceleration enhancements. Such improvements would help to enable any of the major long-term alternatives for powering vehicles in the future without favoring one technology (e.g., hydrogen) over any other (e.g., biofuels). Vehicle performance standards could include a requirement that some number of vehicles be zero or near-zero emitting, which would favor biofuels, batteries, or hydrogen vehicles over conventional gasoline vehicles, but leave it to the market to determine the most viable option. Such a requirement could be pegged, for example, to the hydrogen vehicle penetration rates achieved by HFCV-specific policies described above; this would guarantee that these vehicles were low-emitting, but not necessarily hydrogen fuel cell vehicles.

Another policy option is to raise gasoline taxes. Increasing per-mile costs of driving would discourage unnecessary trips and promote fuel economy and alternative fuel vehicles, including hydrogen. Thus, emissions and oil use could be reduced throughout the transportation sector. However, there has been considerable opposition to higher fuel taxes in the past.

Market-based greenhouse gas control instruments—namely, cap-and-trade or CO_2 taxes—are also being implemented or under consideration both here and abroad. If these instruments are applied economy-wide, they can effectively exploit all low-cost emission reduction opportunities. However, for reasons already discussed, these policies most likely would not provide sufficient pricing incentives by themselves to speed the adoption of major transformational technologies, such as fuel cell vehicles. If carbon emission allowances are allocated through an auction, cap-and-trade and tax proposals under consideration in the United States could generate \$50 billion to \$100 billion or more worth of annual allowances or government revenue. Many of these legislative proposals set aside allowances or government revenue to fund technology programs or incentives, providing a potential stable funding source for the types of technology programs discussed earlier.

CONCLUSIONS

CONCLUSION: Sustained, substantial, and aggressive energy security and environmental policy interventions will be needed to ensure marketplace success for oil-saving and greenhouse-gas-reducing technologies, including hydrogen fuel cell vehicles.

CONCLUSION: Policies designed to accelerate the penetration of HFCVs into the U.S. vehicle market will be required to exploit the long-term potential of HFCVs. The committee concluded that these policies must be durable over the transition time frame but should be structured so that they are tied to technology and market progress, with any subsidies phased out over time. Such policies are likely to deliver significant long-term reductions in U.S. oil demand, but additional policies limiting greenhouse gas emissions will be required in order to also reduce CO_2 emissions significantly.

If policy makers decide to push the market penetration of HFCVs, these policies should be seen as a complement to (not a substitute for) broader measures to reduce greenhouse gases and oil use. However, the committee believes that HFCV-push policies would have to be designed very carefully if they are to be effective. Incentives must be substantial, durable, and coordinated with measures to encourage the development of a hydrogen fuel distribution infrastructure and continued basic research on vehicle design. Moreover, it is not entirely clear whether a price-based (subsidy) approach or a quantity-based (quota) approach to promoting vehicle sales is preferable; both approaches have pros and cons. A mix of the two approaches, however, might afford the most effective policy to achieve the maximum practicable deployment of fuel cell vehicles in the time frame examined in this study.

9
Advantages and Disadvantages of a Transition to Hydrogen Vehicles in Accordance with the Time Lines Established by the Budget Roadmap

The transition to hydrogen vehicles in accordance with the time line established by the budget roadmap will have specific advantages and disadvantages. This chapter addresses those considerations. Throughout this chapter, it is assumed that the transition follows the maximum practicable scenario identified by the committee and discussed in detail in Chapters 6 and 7: Case 1, Hydrogen Success. The reader should keep in mind that the committee considers that this scenario represents the fastest possible transition to a significant number of hydrogen fuel cell vehicles (HFCVs), and that the numbers and timing discussed in this chapter are not to be viewed as the committee's projections of what will *necessarily* happen.

The reader should also keep in mind what this chapter is not. First, this chapter does not address the risks of technical shortfalls or failures themselves. These "technical" risks are addressed via the various cases analyzed and discussed in Chapter 6, especially via Case 1b, Hydrogen Partial Success. Instead this chapter focuses on the broader advantages and disadvantages of the transition, *assuming* it is accomplished in accordance with the time line and budget roadmap of Case 1.

Second, the reader should note that this chapter also does not present a discussion of alternative technical or public policy approaches to meeting the twin policy goals of reduction of oil imports and of carbon dioxide (CO_2) emissions. As with the risks of technical shortfall or failure above, some alternative approaches (efficiency improvements, biofuels) are analyzed in detail and discussed in the various cases presented in Chapter 6, especially Cases 2 and 3. Other alternative approaches have been discussed briefly, without detailed analysis, in various sections throughout this report and recommended for further study if the committee regarded this as appropriate. Again, as above, the reader should keep in mind that this chapter focuses only on the broader advantages and disadvantages of the transition itself, *assuming* it is accomplished in accordance with the time line and budget roadmap of Case 1.

ANTICIPATED BENEFITS AND COSTS OF THE TRANSITION

As explained in Chapter 2, a successful transition to hydrogen fuel cell vehicles, or a successful transition to any revolutionary new personal transportation system, would be a long-term undertaking facing both significant technical and market risks whose details are impossible to predict. One general statement can be made, however: the anticipated benefits are almost all long term and strategic in nature, while the required investments begin (or continue) immediately and must be sustained for many years before their benefits are manifest. This fact creates a natural tension between short-term costs and long-term benefits that must be addressed if the United States is to make such a transition. Historically, federal funds have been used to lessen or overcome the tensions in dealing with such long-term investments (see, for example, Griliches, 1960; Mansfield, 1966). Chapter 8 of this report discusses some federal government actions that might facilitate the transition to hydrogen fuel cell vehicles.

Most of this chapter relies on the committee's analysis to identify the advantages or benefits that would result from a successful transition to HFCVs and to set out the costs or disadvantages that could be expected. It also discusses briefly and recommends for further study some broader potential benefits and risks that may be attributable to an HFCV transition, but whose detailed analysis was outside this committee's scope.

Advantages or Benefits

The primary benefits expected from a successful transition to a hydrogen-based transportation system are captured as two major policy goals that formed the basis for this study:

1. Reductions in imports of oil and
2. Lower CO_2 and other greenhouse gas emissions.

The committee also notes that additional benefits may accrue (e.g., public health benefits from reduction in air pollution) from this transition; such other benefits are discussed briefly in this chapter. However, the two primary benefits above, called out in the statement of task, were the focus of the committee's effort.

Reduced Oil Consumption

It is difficult for the U.S. oil industry to increase domestic oil production due to declining production from existing oil fields, environmentally restricted acreage, and the complexity of new exploration and production projects, especially offshore. Therefore any significant reduction of imports probably would require a concomitant reduction in demand for oil. Reduction of oil imports offers two main benefits to the United States:

1. Improved energy security, at least to the extent that reduced oil imports are accompanied by the development and adoption of a more diverse set of indigenous energy sources for U.S. transportation, such as coal, nuclear power, biofuels, or other renewable resources; and
2. Potential for long-term reduction of the outflow of dollars currently required to pay for the nation's energy needs, especially as indigenous sources of energy are eventually exploited to produce hydrogen. It is also possible that decreased pressure on world oil markets may contribute to a reduction in the price of the oil that must still be imported.

The U.S. transportation sector consumed 28 quadrillion British thermal units (Btu) (28 quads) of energy in 2006, representing 28 percent of total energy consumed. Furthermore, 96 percent of the energy used in the transportation sector was consumed in the form of petroleum products (DOE-EIA, 2007, Tables 2.1a and 2.1e). Furthermore, in 2006, about two-thirds of the crude oil used in the United States was imported (12.3 million barrels per day out of a total of 20.6 million barrels per day, or approximately 60 percent), a proportion that has grown steadily since the early 1980s (DOE-EIA, 2007, Diagram 2 and Figure 5.1).

As shown in Figure 6.32 in Chapter 6, the alternative approaches studied by the committee (internal combustion engine [ICE] improvements and biofuels) offer significant reductions in oil consumption by 2020, but HFCVs are on the path to achieve much more significant savings in the 2035-2050 time frame, at a time when the rate of improvement in oil import reduction due to biofuels and ICE improvements would be slowing.

A further benefit (although not unique) of the use of hydrogen as a transportation fuel is the multiplicity of fuel resources and production methods from which hydrogen can be made, including distributed and central-station steam methane reformers (SMRs) used to convert natural gas to hydrogen, coal gasification, biomass gasification, and electrolysis of water (using grid electricity, renewable energy, or nuclear power; see Table 6.1). Although those fuels and pathways that rely more heavily on indigenous U.S. energy resources (e.g., coal gasification, biomass gasification, and water electrolysis with renewable or nuclear power) today require additional development, all represent alternatives that might be able to mitigate the impact of a significant disruption in the availability of crude oil or natural gas imports.

Reductions in CO_2 Emissions

As shown in Figure 6.33, the alternative technologies reviewed by the committee—(1) evolutionary efficiency improvements to vehicles with internal combustion engines and (2) biofuels—have the potential to achieve significant reductions in greenhouse gas emissions by 2020. The former has been incorporated in the reference case until 2020 and could continue to improve efficiency thereafter. However, one can also see in Figure 6.33 that growth in the benefits from these alternative technologies could slow significantly in subsequent years under the scenarios used in this study, while the benefits from adoption of HFCVs, whose numbers begin to be significant in the 2020-2025 time frame, are on a path to increase rapidly throughout 2035-2050 under the maximum practicable scenario. Although it is difficult to predict many years into the future, the sense of the committee is that these trends seem reasonable: the impact of biofuels in the United States is limited by available land and water, and improvements to ICE vehicles are limited by considerations such as cost, how much more efficient engines can be while still meeting durability and environmental requirements, and how much weight can be removed from the vehicle while still meeting consumer preferences. During that same period, the benefits from HFCVs have the potential to continue growing, due both to technology improvements in these relatively new systems and to increasing market penetration. Thus, a transition to HFCVs offers the potential, if successful, to eventually achieve benefits exceeding those of the alternative technologies.

Finally, it should be noted that simply transitioning to hydrogen fuel cell vehicles will not necessarily result in the magnitude of CO_2 reductions shown here. Those reductions will depend on the pathways via which hydrogen is produced, as well as on the higher efficiency of HFCVs relative to conventional gasoline engines. As noted in Chapter 6, during the transition period when hydrogen is assumed to be produced via reforming of natural gas, the life-cycle greenhouse gas emissions of HFCVs are still lower than those of conventional vehicles, thanks largely to the much higher efficiency of fuel cells. In the longer term, after about 2025, hydrogen is assumed to be supplied increasingly from central coal-based plants with carbon capture and sequestration (CCS). As noted in earlier chapters of this report, strong policy drivers limiting CO_2 emissions will be required to implement CCS at central coal plants. To the extent that CCS technology proves too difficult or too expensive to realize, emissions of CO_2 would increase relative to the values in Figure 6.33 unless other options for low-carbon hydrogen production were used.

Costs and Risks of a Transition to Hydrogen Fuel Cell Vehicles

A transition to hydrogen fuel cell vehicles will also have substantial costs as well as various potential risks, as discussed below.

Costs

As discussed in Chapter 7, sustained expenditures are required for a successful transition to HFCVs, initially in support of technology research, development, and demonstration (RD&D) programs, and later to support the initial construction of the hydrogen infrastructure and the introduction of HFCVs into the market. Estimated expenditures for the public investment alone are roughly $55 billion over the 16 years from 2008 to 2023 for the Hydrogen Success scenario.

It is important to put this estimate into perspective. First, U.S. consumers are going to spend more than $7 trillion on new vehicles and at least $4.5 trillion on fuel over the 16-year period. The large auto manufacturers spent a total of $38 billion in 2006 on RD&D (in all areas, not just for hydrogen), and the combined capital budgets of the three largest integrated energy companies exceeded $20 billion in the same year.[1] Considering another energy subsidy program, the recently passed U.S. energy bill with ethanol mandates, will result in more than $160 billion in subsidies to the ethanol industry over the next 16 years, assuming the subsidies are extended though that time frame.[2] Although the committee clearly understands that none of these funding numbers are truly comparable from an investment and risk standpoint, they do help frame the discussion about the magnitude of the possible hydrogen expenditure levels.

Furthermore, since these investments occur over time, some of the risks can be mitigated by periodic assessment of both the progress of various technologies and the current environment for development, and subsequent rebalancing of the portfolio of programs and development activities based on these assessments. It should be noted that, as shown in Figure 7.3, the rate of public investment remains moderate during technology development and demonstration, through approximately 2012, and accelerates significantly only when policy options are required to facilitate commercial introduction of fuel cell vehicles to the market and wider rollout of early hydrogen fuel infrastructure.

Indirect costs may accrue as well, such as the loss of tax revenue to governments as sales of (presumed) tax-free hydrogen substitute for sales of taxed gasoline. A detailed analysis of the impact of such potential indirect costs was beyond the scope of this study.

Finally, the committee notes that the expenditures (and the use of other key resources such as skilled manpower) made to further a transition to hydrogen fuel cell vehicles also incur real, but somewhat hidden, "opportunity costs"—that is, as a result of being spent on the HFCV transition, these funds are not available for any other purpose, so other opportunities for the use of the funds are forgone (Economist, 2007). For example, in a narrow sense, a premature and too-specific focus only on HFCVs might divert resources away from other alternatives with potential benefits in terms of reduced oil imports and CO_2 emissions, such as biofuels, efficiency, batteries, or hybrids. In a broader sense, HFCV expenditures simply may be regarded as diverting funds away from any other program with potential public benefit. An analysis of all potential opportunities for these funds is a matter of public policy, however; and thus any detailed analysis of the opportunity costs associated with this transition is also beyond the scope of this study.

Risks

The committee has identified three types of potential risks associated with the time line and budget roadmap established for the transition to HFCVs. All of these would result in opportunity costs, financial losses, or failure to achieve expected reductions in oil use and/or CO_2 emissions if the transition is not successful:

1. Potentially limited market acceptance,
2. Difficulty of achieving simultaneous transitions of vehicles and fuel infrastructure, and
3. Reliance on geological sequestration to mitigate CO_2 emissions from hydrocarbon-based hydrogen production.

Limited Market Acceptance. Although several vehicle manufacturers have established detailed demonstration and product development time lines for fuel cell vehicles, including multiple rounds of prototypes, field tests, and consumer acceptance activities, the potential remains for market acceptance to take longer than, or sales volumes to fall short of, the committee's projections. Slow growth might occur owing to problems in achieving technology goals, issues with fuel supply or vehicle resale values, customer perceptions of hydrogen safety, safety concerns expressed in local zoning

[1]Data assembled by the committee from the websites of the seven largest auto companies: GM, Ford, Chrysler, Toyota, Honda, Nissan, and Volkswagen.

[2]The Energy Security and Independence Act of 2007 increased the required annual volume of renewable fuel to 30 billion gallons by 2020 and 36 billion by 2022. The credit remains the same at $0.51 per gallon. The $160 billion figure was calculated by assuming a reasonable growth curve to meet these production goals with a constant tax credit over the 16 years. No discount rate or adjustment for inflation was applied.

codes, or other factors. No matter the root cause, resulting impacts would likely include the following:

- Poor returns on investments, either when development programs are dragged out too long or sales do not occur in a timely fashion;
- An unnecessary drag on the U.S. economy due to underutilized or stranded installed base; and
- Risk of suboptimal technology choices, if these choices were forced before the markets were ready or if a superior alternative becomes available, such as greatly improved batteries that permit extended-range electric vehicles.

Difficulty of Achieving Simultaneous Transitions of Vehicles and Fuel Infrastructure. Simultaneously carrying out a transition in the fuel infrastructure of light-duty vehicles and a transition in the type of light-duty vehicles being driven represents a challenge not faced before by the United States. Without sales of fuel cell vehicles, fuel providers will be reluctant to invest in fueling capability; without both actual and perceived fueling capability (convenient station locations, fueling speed, and safety), consumers will be reluctant to purchase fuel cell vehicles. However, it should be noted that the committee estimates that the infrastructure transition costs will be comparable to other costs that industry currently manages.

Risks of Reliance on Carbon Sequestration. If hydrogen is going to be made from fossil fuels, as the scenarios in Chapter 6 suggest is likely to be the case for several decades beyond the transition, significant amounts of carbon dioxide captured as part of the hydrogen production process will be emitted to the atmosphere from the production process unless it is sequestered. As discussed in Chapter 3, the most promising option for sequestration is to inject captured CO_2 into deep geological formations where it is expected to remain indefinitely. Although there are substantial ongoing RD&D efforts on carbon sequestration in the United States, it remains an unproven technology for the types and scales of applications envisioned here. Pending a successful outcome of ongoing programs to develop and demonstrate the viability of CCS, and the development of a regulatory structure for such projects, there remain uncertainties and associated risks with assuming that this technology will be available and effective when needed.

OTHER POTENTIAL BENEFITS

The committee notes that other benefits may also accrue from a successful HFCV transition. These benefits may include the following:

- Potential benefits for public health—both directly via reductions of local emissions of criteria pollutants from light-duty vehicles and indirectly via potential mitigation of global warming and its possible detrimental effects on public health;
- Potential economic benefits for the United States if onshore individuals, entrepreneurial companies, or large industrial companies develop and can capture the rents (operating profits, licensing fees, royalties, etc.) from hydrogen technologies; and
- Specific (although perhaps intangible) benefits for particular segments of the consumer light-duty vehicle market, such as environmental friendliness or peace of mind about future fuel availability.

None of these other potential benefits were studied by the committee, but they could be significant and worthy of investigation. Many of these might be realized as well by approaches other than hydrogen fuel cell vehicles.

OTHER POTENTIAL RISKS

The committee also notes that other indirect risks may result from a transition to hydrogen fuel cell vehicles. One area of potential concern that the committee has indentified is potential price pressure on commodities due to increased demand, including but not limited to natural gas, platinum, and food staples, either via direct competition for food stocks as process inputs (e.g., corn for ethanol) or indirect competition for the land, water, and other requirements to produce food stocks. A detailed analysis of these or any other risks requires additional study.

CONCLUSION

CONCLUSION: A portfolio of technologies including hydrogen fuel cell vehicles, improved efficiency of conventional vehicles, hybrids, and use of biofuels—in conjunction with required new policy drivers—has the potential to nearly eliminate gasoline use in light-duty vehicles by the middle of this century, while reducing fleet greenhouse gas emissions to less than 20 percent of current levels. This portfolio approach provides a hedge against potential shortfalls in any one technological approach and improves the probability that the United States can meet its energy and environmental goals. Other technologies also may hold promise as part of a portfolio, but further study is required to assess their potential impacts.

As discussed above, it is not possible to predict the detailed nature of the transition or even whether better alternatives might emerge during the time it takes to accomplish the transition. It will be important for the federal government to adopt policy initiatives that are both ***substantial and durable,*** so that companies—both large and small—can respond to clear market signals.

REFERENCES

DOE-EIA (Department of Energy-Energy Information Agency). 2007. *Annual Energy Review 2006.* Pub. No. DOE/EIA-0384(2006). Available at http://tonto.eia.doe.gov/ merquery/mer_data.asp?table=T02.05, (accessed November 19, 2007).

Economist. 2007. Opportunity Cost. *The Economist*, December; available at http://www.economist.com/research/Economics/alphabetic.cfm?TERM=OECD#opportunitycost.

Griliches, Z. 1960. Hybrid Corn and the Economics of Innovation. *Science* 132 (July 29):275-280.

Mansfield, E. 1966. Technological Change: Measurement, Determinants, and Diffusion. *Report to the President by the National Commission on Technology, Automation, and Economic Progress*, Appendix. Vol. II. Washington, D.C.

Appendixes

Appendix A

Committee Biographical Information

CHAIR

Michael P. Ramage, NAE, is a retired vice president, ExxonMobil Research and Engineering Company. Previously he was executive vice president and chief technology officer, Mobil Oil Corporation. Dr. Ramage held a number of positions at Mobil including research associate, manager of process research and development, general manager of exploration and producing research and technical service, vice president of engineering, and president of Mobil Technology Company. He has broad experience in many aspects of the petroleum and chemical industries. He has served on a number of university visiting committees and was a member of the Government-University-Industry Research Roundtable. He was a director of the American Institute of Chemical Engineers and is a member of several professional organizations. Dr. Ramage chaired the committee that authored the recent National Research Council (NRC) report *The Hydrogen Economy: Opportunities, Costs, Barriers, and Research Needs*. He is a member of the National Academy of Engineering (NAE) and has served on the NAE Council. Dr. Ramage has B.S., M.S., Ph.D., and HDR degrees in chemical engineering from Purdue University.

MEMBERS

Rakesh Agrawal, NAE, is Winthrop E. Stone Distinguished Professor, School of Chemical Engineering, Purdue University. Previously, he was an Air Products fellow at Air Products and Chemicals, Inc., where he worked from 1980 to 2004. A major thrust of his research is related to energy issues and includes novel processes for fabrication of low-cost solar cells, biomass and coal-to-liquid fuel conversion, hydrogen production from renewable sources, and energy systems analysis. His research interests further include basic and applied research in gas separations, process development, synthesis of distillation column configurations, adsorption and membrane separation processes, novel separation processes, gas liquefaction processes, cryogenics, and thermodynamics. He holds 116 U.S. and more than 500 foreign patents. These patents are used in more than 100 chemical plants with a capital expenditure in excess of a billion dollars. He has authored 66 technical papers and given many lectures and presentations. He chaired the Separations Division and the Chemical Technology Operating Council of the American Institute of Chemical Engineers (AIChE) and also a Gordon Conference on Separations. He was a member of the NRC Committee on Alternatives and Strategies for Future Hydrogen Production and Use. He is currently a member of the AIChE's Board of Directors and its Energy Commission. He is also a member of the NRC Board on Energy and Environmental Systems (BEES). He has received several awards, including the J&E Hall Gold Medal from the Institute of Refrigeration (UK) and the Presidential Citation for Outstanding Achievement from the University of Delaware. The AIChE has honored him with the Clarence G. Gerhold Award, the Institute Award for Excellence in Industrial Gases Technology, the Chemical Engineering Practice Award, and chose him to give the Institute Lecture at the 2005 AIChE Annual Meeting. Dr. Agrawal received a B. Tech. from the Indian Institute of Technology, in Kanpur, India; an M.Ch.E. from the University of Delaware; and an Sc.D. in chemical engineering from the Massachusetts Institute of Technology.

David L. Bodde serves as a professor and senior fellow at Clemson University. There, he directs innovation and strategy at Clemson's International Center for Automotive Research. Prior to joining Clemson, Dr. Bodde held the Charles N. Kimball Chair in Technology and Innovation at the University of Missouri in Kansas City. Dr. Bodde serves on the boards of directors of several energy and technology companies, including Great Plains Energy and the Commerce Funds. His executive experience includes vice president, Midwest Research Institute; assistant director of the Congressional Budget Office; and deputy assistant secretary in the U.S. Department of Energy. Dr. Bodde frequently testifies before congressional committees. He was once a soldier and served in the Army in Vietnam. He has a doctorate in business administration from Harvard University, M.S. degrees in nuclear engineering (1972) and management (1973), and a B.S. from the United States Military Academy.

David Friedman is research director, Clean Vehicles Program, Union of Concerned Scientists (UCS), Washington, D.C. He is the author or coauthor of more than 30 technical papers and reports on advancements in conventional, fuel cell, and hybrid electric vehicles and alternative energy sources, with an emphasis on clean and efficient technologies. Before joining UCS in 2001, he worked for the University of California-Davis (UC Davis) in the Fuel Cell Vehicle Modeling Program, developing simulation tools to evaluate fuel cell technology for automotive applications. He worked on the UC Davis FutureCar team to build a hybrid electric family car that doubled its fuel economy. He previously worked at Arthur D. Little researching fuel cell, battery electric, and hybrid electric vehicle technologies, as well as photovoltaics. He served as a member of the NRC Panel on the Benefits of Fuel Cell R&D of the Committee on Prospective Benefits of the U.S. Department of Energy's (DOE's) Energy Efficiency and Fossil Energy R&D Programs, Phase 1; on the Panel on Benefits of DOE's Light-Duty Hybrid Vehicle R&D Program; and as a member of the NRC Committee on National Tire Efficiency. He earned a bachelor's degree in mechanical engineering from Worcester Polytechnic Institute and is a doctoral candidate (2007) in transportation technology and policy at UC Davis.

Susan Fuhs is president, Conundrum Consulting. Previous positions include general manager, Astro Aerospace; general manager, GE Hybrid Power Generation Systems; director, New Ventures, Honeywell International; technology policy analyst, RAND; and project engineer, Advanced Applications, AlliedSignal Aerospace. Dr. Fuhs' technical and business experience has focused on overcoming barriers to the development and implementation of advanced technologies. Her experience with fuel cells includes developing fuel cell systems for stationary and transportation applications, including fuel cells for the Partnership for a New Generation of Vehicles; developing fuel cell marketing and business plans; and managing the solid oxide fuel cell subsidiary of General Electric Power Systems. She currently consults in strategic planning, new product development, business development, and technology roadmapping. She is a past board member, National Hydrogen Association, and past chairperson, Space Systems Technical Committee, American Institute of Aeronautics and Astronautics. She has a Ph.D. and an M.S. in mechanical engineering and a B.S. in chemical engineering from the California Institute of Technology, and an M.B.A. from the Anderson School, University of California, Los Angeles.

Judi Greenwald is the director of Innovative Solutions at the Pew Center for Global Climate Change. She oversees the solutions program and develops mechanisms for learning about and promoting innovative solutions—including research, publications, web-based information and databases, and workshops. Ms. Greenwald focuses on technological innovation, business solutions, and state and regional solutions.

Ms. Greenwald has more than 20 years of experience working on energy and environmental policy. Prior to coming to the Pew Center, she worked as a consultant, focusing on innovative approaches to solving environmental problems, including climate change. She also served as a senior adviser on the White House Climate Change Task Force. As a member of the professional staff of the U.S. Congress Energy and Commerce Committee, she worked on the 1990 Clean Air Act Amendments, the 1992 Energy Policy Act, and a number of other energy and environmental statutes. She was also a congressional fellow with then-Senate Majority Leader Robert C. Byrd, an environmental scientist with the U.S. Nuclear Regulatory Commission, and an environmental engineer and policy analyst at the U.S. Environmental Protection Agency (EPA).

Ms. Greenwald has a bachelor of science in engineering, cum laude, from Princeton University, and an M.A. in science, technology, and public policy from George Washington University.

Ms. Greenwald has published papers on the future of water quality monitoring, worker and community adjustment to climate change policy, a multimedia approach to radon, environmental policies affecting the development of newer coal technologies, and the implications for air quality analysis of extended lifetimes for coal-fired boilers.

Robert L. Hirsch is senior energy adviser, Management Information Services, Inc. (MISI). Formerly he was senior energy program adviser at SAIC. His past positions include senior energy analyst with the RAND Corporation; executive adviser to the president of Advanced Power Technologies, Inc.; vice president, Washington Office, Electric Power Research Institute (EPRI); vice president and manager, Research and Technical Services Department, ARCO Oil and Gas Company; chief executive officer of ARCO Power Technologies, a company that he founded; manager, Baytown Research and Development Division, and general manager, Exploratory Research, Exxon Research and Engineering Company; assistant administrator for solar, geothermal, and advanced energy systems (presidential appointment) and director, Division of Magnetic Fusion Energy Research, U.S. Energy Research and Development Administration. He has served on numerous advisory committees including as a member of the DOE Energy Research Advisory Board and a number of DOE national laboratory advisory boards. He has served on several NRC committees, including the one that authored the report *Fuels to Drive Our Future*, which examined the economics and technologies for producing transportation fuels from U.S. domestic resources, the Committee on Alternatives and Strategies for Future Hydrogen Production and Use, and as chair of the Committee to Examine the Research Needs of the Advanced Extraction and Process Technology Program. He served as chairman of the Board on Energy and Environmental Systems and is a national associate of the Academies. He brings expertise in a number of areas of science and technology and business related to energy production and consumption, research and development, and public policy. He received a Ph.D. in engineering and physics from the University of Illinois.

James R. Katzer, NAE, is an independent consultant. He has recently been a visiting scholar at the Massachusetts Institute of Technology (MIT) working on an MIT study "The Future of Coal in a Carbon Constrained World." Prior to that he was manager of strategic planning and program analysis for ExxonMobil Research and Engineering Company, where he was responsible for technology planning and analysis activities. Before that he was vice president, Technology, Mobil Oil Corporation, with primary responsibilities for ensuring Mobil Oil overall technical health, developing forward-looking technology scenarios, identifying and analyzing technology and environmental developments and trends, and identifying future threats and opportunities and strategies to deal with them. Dr. Katzer joined the Central Research Laboratory of Mobil Oil Corporation in 1981, later becoming manager of process research and technical service and vice president of planning and finance for Mobil Research and Development Corporation. Before joining Mobil he was a professor on the chemical engineering faculty at the University of Delaware and the first director of the Center for Catalytic Science and Technology there. He recently served on the NRC Committee on Alternatives to Indian Point that evaluated various energy supply and end use technologies as potential replacements for the Indian Point nuclear power plants. Dr. Katzer has more than 80 publications in technical journals, holds several patents, and coauthored and edited several books. He received a B.S. degree from Iowa State and a Ph.D. in chemical engineering from MIT.

Gene Nemanich is the retired vice president, Hydrogen Systems, for Chevron Technology Ventures where he was responsible for hydrogen supply and developing and commercializing new hydrogen technologies. He has 32 years of experience with integrated oil companies, including Exxon, Cities Service, Texaco, and Chevron. He has also worked in the areas of refining, clean coal technology, oil supply and trading, and research leading to the development of new hydrogen systems. He represented Texaco in the California Fuel Cell Partnership in 2000-2001, and was a director of Texaco Ovonic Hydrogen Systems LLC, a joint venture with Energy Conversion Devices to commercialize metal hydride hydrogen storage systems. He was one of seven industry leaders that helped prepare the DOE-sponsored Hydrogen Roadmap and has served as chairman of the National Hydrogen Association. He has a B.S. in chemical engineering from the University of Illinois and an M.B.A. from the University of Houston.

Joan Ogden is professor of environmental science and policy and energy policy analyst at the Institute of Transportation Studies, University of California, Davis. Previous to this, she held a number of positions at various research institutions including research scientist, Center for Energy and Environmental Studies, Princeton University. Most of her work has involved technical and economic assessments of new energy technologies, including renewable fuels, the use of hydrogen as an energy carrier, and applications of fuel cell technology in transportation. Particular areas of interest are the production of renewable fuels, the use of hydrogen as an energy carrier, and applications of fuel cells in transportation and stationary power production. Over the past decade, Dr. Ogden has carried out a series of assessments of fuel cell vehicles and hydrogen refueling infrastructure. For 2 years, she served as chairman of the Solar Fuels and Transportation Division of the American Solar Energy Society. She has worked with the H2A, a group of hydrogen analysts convened by the Department of Energy to develop a consistent framework for analyzing hydrogen systems and, in 2005 and 2006, received R&D Excellence awards from the DOE for her work with H2A. In 2004, Dr. Ogden served on the Governor of California's Advisory Panel developing a blueprint plan for the proposed California Hydrogen Highway Network. Dr. Ogden has published more than 100 technical articles on energy topics and one book *Solar Hydrogen*. She received her Ph.D. in physics from the University of Maryland and a B.S. in mathematics from the University of Illinois, Champaign-Urbana.

Lawrence T. Papay, NAE, is currently a consultant with a variety of clients in electric power and other energy areas. His previous positions include senior vice president for the Integrated Solutions Sector, SAIC; and senior vice president and general manager of Bechtel Technology and Consulting. He also held several positions at Southern California Edison, including senior vice president, vice president, general superintendent, and director of research and development (R&D), with responsibilities for areas including bulk power generation, system planning, nuclear power, environmental operations, and development of the organization and plans for the company's R&D efforts. His professional affiliations have included the EPRI Research Advisory Committee, the Atomic Industrial Forum, the DOE Energy Research Advisory Board, and the Renewable Energy Institute. He is a member of the National Academy of Engineering and the National Science Foundation's Industrial Panel on Science and Technology. His expertise and knowledge ranges across a wide variety of electric system technologies, from production, to transmission and distribution, utility management and systems, and end use technologies. He received a B.S. in physics from Fordham University, and an S.M. and Sc.D. in nuclear engineering from MIT.

Ian W.H. Parry is a senior fellow at Resources for the Future. Previous positions include adjunct professor, Department of Economics, Georgetown University; research fellow, U.S. Department of Agriculture; professor, Center for Economic Research and Graduate Education (Prague); and lecturer, Department of Economics, Australian National University. Dr. Parry's research focuses primarily on environmental, transportation, tax, and public health policies. His recent work has analyzed gasoline taxes, fuel economy standards, transit subsidies, alcohol taxes, policies to reduce traffic congestion and accidents, environmental tax shifts, the role of technology policy in environmental protection, the incidence of pollution control policies, and the interactions between regulatory policies and the broader tax system. He received a Ph.D. in economics from the University of Chicago, an M.A. in economics from Warwick University, and a B.A. in economics from the University of Sheffield.

William F. Powers, NAE, is retired vice president, Research, Ford Motor Company. His approximately 20 years at Ford included positions as director, Vehicle, Powertrain and Systems Research; director, Product and Manufacturing Systems; program manager, Specialty Car Programs; and executive director, Ford Research Laboratory and Information Technology. Prior positions also include professor, Department of Aerospace Engineering, University of Michigan, during which time he consulted with the National Aeronautics and Space Administration (NASA), Northrop, Caterpillar, and Ford; research engineer, University of Texas; and mathematician and aerospace engineer, NASA Marshall Space Flight Center. Dr. Powers is a fellow of the Institute of Electrical and Electronics Engineers, Society of Automotive Engineers, American Society of Mechanical Engineers, and International Federation of Automatic Control. He is a member of the National Academy of Engineering and a foreign member of the Royal Swedish Academy of Engineering Sciences. He has extensive expertise in advanced research and development of automotive technology. He is a member of the National Academies Board on Energy and Environmental Systems and recently served on the Committee on Alternatives and Strategies for Future Hydrogen Production and Use. He has a B.S. in aerospace engineering, University of Florida, and a Ph.D. in engineering mechanics, University of Texas-Austin.

Edward S. Rubin is the Alumni Professor of Environmental Engineering and Science at Carnegie Mellon University (CMU). He holds joint appointments in the Departments of Engineering and Public Policy and Mechanical Engineering and is the founding director of CMU's Environmental Institute and Center for Energy and Environmental Studies. His teaching and research interests at CMU are in the areas of environmental control, energy utilization, and technology-policy interactions, with a particular focus on coal-based systems. His expertise includes modeling and assessment of energy and environmental systems with applications to electric power generation technologies, energy use, and emission control systems; global climate change policy issues; carbon sequestration and management; and environmental technology innovation and its relation to government policies. He has served as a member of numerous technical and advisory committees, including to the U.S. Environmental Protection Agency, U.S. Department of Energy, Intergovernmental Panel on Climate Change, and the National Academy of Sciences-National Research Council. He is a past chairman of the Environmental Control Division of the American Society of Mechanical Engineers. He earned a B.E. in mechanical engineering at the City College of New York and an M.S. and Ph.D. in mechanical engineering from Stanford University.

Robert W. Shaw, Jr., is president of Aretê Corporation, the manager of the Micro-Generation Technology Fund, LLC, and the five Utech venture funds. He has more than 20 years of experience in the venture capital industry and is a leader in developing modular-dispersed generation, renewable energy generation, hydrogen energy systems, and specialty materials. He previously held the position of senior vice president and member of the Board of Directors of Booz, Allen & Hamilton, where he was a founder of the firm's Energy Division, which provided management and technical consulting services to utilities and energy companies. He also held research positions at Bell Laboratories and Cavendish Laboratory directed at the electronic and structural properties of materials. Dr. Shaw served for 11 years as director and chairman of Distributed Energy Systems Corporation (DESC) and for 5 years as director and chairman of CTP Hydrogen Corporation. He has been a director of H2Gen Innovations, Inc., since 2001. He has served as a member of the NRC Board on Energy and Environmental Systems and the Panel on Benefits of DOE's Fuel Cell R&D Program. He is also a member of DOE's Hydrogen Technology Advisory Committee. He has a Ph.D. in applied physics from Stanford University, an M.S.E.E. from Cornell University, and an MPA in organization design from American University.

Arnold F. Stancell, NAE, is Turner Professor of Chemical and Biomolecular Engineering, emeritus, Georgia Institute of Technology. He has also served as visiting associate professor and visiting professor of chemical engineering, MIT. Prior to joining Georgia Tech, he held various research positions and subsequently a number of management positions at Mobil Oil Corporation including vice president, International Exploration and Production, vice president, worldwide Marketing and Refining Planning, and regional executive in Europe. He left Mobil in 1993 after a 31-year career including 10 years in research where he was awarded nine patents and worked on innovative processes for various hydrocarbon processes and processes for permselective membranes for separations applications. His research interests are in polymer and petrochemical processes and plasma reactions in microelectronics processing. His technical and business experience, and economic analysis experience for large projects, bring a perspective on infrastructure costs and requirements to this study. He is a member of the National Academy of Engineering and received the AIChE National Award in Chemical Engineering Practice. Other awards include the Outstanding Teacher Award in Chemical Engineering, Georgia Tech (1997), Career Achievement Award of City College of New York (1993), Black Engineer of the Year (1992), Invited Marshall Lecturer, University of Wisconsin (1991), and Professional Achievement Award of the National Organization for the Professional Advancement of Black Chemists and Chemical Engineers (NOBCChE) (1975). He has an Sc.D. from MIT (1962) and a B.S. (magna cum laude, 1958) from City College of the City University of New York, in chemical engineering.

Tony Wu is senior research engineer and project manager, Fuel Cell, Hydrogen, Electric Transportation, and Energy Storage, Southern Company. Previous positions include staff technology engineer, Energizer Power Systems, and technology engineer, Gates Energy Products. He has more than 15 years of combined experience in fundamental research, technology development and product testing and validation, and project management. His primary technical expertise is in chemical, material, and electrochemical behavior of various electrochemical systems including ultracapacitor, battery, and fuel cell technologies. At Southern Company, he is directly responsible for multiple research program areas, including distributed energy resources, hydrogen, electric transportation, and energy storage programs. His research activities focus on technical evaluation of various technologies, determining the applicability of these technologies in both stationary and transportation applications, and making recommendation to management for proper strategic actions. He has a B.S. in chemical engineering from Tamkang University (Taiwan) and an M.S. in chemical engineering from Auburn University.

Appendix B

Presentations at Committee Meetings

FIRST COMMITTEE MEETING: FEBRUARY 20-22, 2007, WASHINGTON, D.C.

JoAnn Milliken, Program Manager, Hydrogen, Fuel Cells, and Infrastructure Technologies, U.S. Department of Energy (DOE): Overview of the Hydrogen Challenge and Federal Effort

Sig Gronich, Office of Hydrogen, Fuel Cells and Infrastructure Technologies Program, DOE: Overview of DOE's Transition Analysis Effort

Robert Rose, U.S. Fuel Cell Council: Discussion of Fuel Cell Council Report, Fuel Cells and Hydrogen: The Path Forward

Arthur Katsaros, Air Products: Strategies and Costs for a Hydrogen Fuel Infrastructure

Dan Rastler, Electric Power Research Institute: Roles and Opportunities for Power Companies in the Hydrogen-electric Economy

Robert W. Shaw, Aretê Corporation: Role of Entrepreneurs and the Venture Capital Sector in a Transition

Joan Ogden, University of California, Davis: Hydrogen Transition Issues

SECOND COMMITTEE MEETING: APRIL 18-20, 2007, WASHINGTON, D.C.

Ben Knight, Honda Motor Corporation: Honda's Hydrogen Programs and Expectations

Phillip Baxley, Shell Hydrogen: Shell's Hydrogen Programs and Expectations

K.G. Duleep, ICF: The Hydrogen Transition and Competing Conventional Fuel Economy Technology

JoAnn Millikin, DOE: DOE's Perspective on How the Committee's Study Could Be Used by DOE's Programs

Fred Joseck and Mark Paster (DOE), Brian James (Directed Technologies Inc.): Meeting with Committee's Infrastructure Team

Richard Hess and Maggie Mann, DOE: Biomass

Lowell Miller, DOE: Coal, FutureGen and CO_2 Sequestration Programs

David Greene, Oak Ridge National Laboratory (ORNL): Modeling, Scenarios, and the Transition

Steve Plotkin, Argonne National Laboratory (ANL): Examining Hydrogen Transitions

Frances Wood (On Location, Inc.), Fred Maples (DOE-Energy Information Agency); Timothy Johnson (U.S. Environmental Protection Agency): Meeting with committee's Scenarios Team

Johanna Levee (National Renewable Energy Laboratory) and Knut Harg (Norsk-Hydro), Brian James (Directed Technologies Inc.), and Mark Paster (DOE): Meeting with Committee's Technology Readiness Team

THIRD COMMITTEE MEETING: JUNE 25-26, 2007, WASHINGTON, D.C.

Kelly Fletcher, GE Research: Hydrogen; Transitioning to Carbon-free Energy

Jon Bereisa, General Motors: GM's Fuel Cell Plans and Programs

Taiyo Kawai and Bill Reinert, Toyota: Toyota's Plans and Programs for Hydrogen and Other Advanced Vehicles

Catherine Dunwoody, California Fuel Cell Partnership: California's Plans for Hydrogen

Bryan Jenkins, University of California, Davis: Biofuels and H_2 from Biomass

Appendix C

Modeling a Hydrogen Transition

Joan Ogden, Marc Melaina, and Chris Yang

A goal of the scenario analysis is to estimate the investments needed to bring hydrogen fuel cell vehicles to life-cycle cost competitiveness with a reference gasoline vehicle. To aid this process, researchers at the University of California, Davis (UC Davis,) developed a relative simple, flexible, transparent EXCEL model called STM (Simple Transition Model) that the committee used to look at how hydrogen transition costs depend on key variables.

Inputs to the model include

- Market penetration rate of hydrogen fuel cell vehicles (HFCVs)
- Cost of HFCVs versus cumulative production, time (learning rate, scale factors for manufacturing HFCVs)
- HFCV performance over time (fuel economy)
- Cost and performance of baseline reference vehicle (gasoline internal combustion engine vehicle [ICEV]) over time
- Oil (gasoline) price over time
- Cost of hydrogen ($/kg) over scale, time
  —Costs and performance for H_2 infrastructure components are included in H2A and UC Davis models
- Source of hydrogen over time and greenhouse gas (GHG) emission factors

Outputs include

- Scenario description
- "Breakeven" year, when HFCVs become competitive with reference ICEVs on a life-cycle cost basis (cost of the vehicle plus the discounted cost of the H_2 to fuel it)
- Transition costs (How much does it cost to get to break even?)
  —Incremental vehicle costs
  —Infrastructure capital costs
  —Policy costs (subsidies, carbon tax, etc.)
- Primary energy use over time
- GHG emissions over time

Figures C.1(a) and C.1(b) show the program's logic and flow, which involves the following five steps.

Step 1: Estimating Infrastructure and Delivered Hydrogen Costs (Figure C.2)

- For each year from 2005 to 2050, the infrastructure needed to serve that H_2 demand is designed using the UC Davis or H2A models.
- The initial H_2 infrastructure is built up in "lighthouse" cities (similar to the Department of Energy [DOE] transition analysis).
- The capital cost for infrastructure is estimated at each time.
- The feedstock and other operating costs are estimated as well.
- This allows determination of the delivered H_2 cost ($/kg) for each year.

Step 2: Cash Flow Analysis: Estimating the Life-cycle Cost (LCC) of Transportation

The life-cycle cost of transportation is estimated for each year (*i* indicates one of these years) from 2005 to 2050 (LCC [i]) for HFCVs compared to what ***would have been paid*** for the same number of reference gasoline vehicles.

NOTE: Joan Ogden is a member of the Committee on Assessment of Resource Needs for Fuel Cell and Hydrogen Technologies. Marc Melaina and Chris Yang worked at the University of California, Davis.

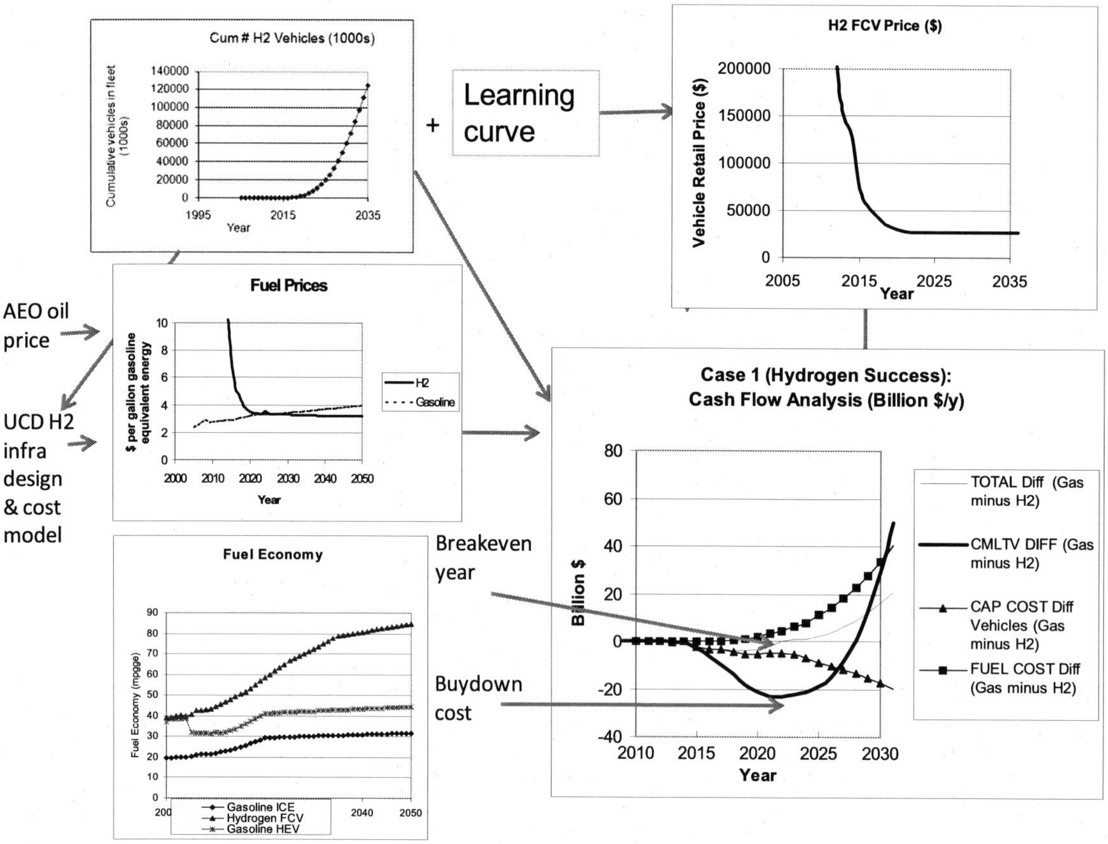

FIGURE C.1(a) Flow diagram of simple transition model (STM) (part 1).

HFCV LCC (i) ($/yr) = number of new HFCVs (in year i) × vehicle first cost (in that year) ($/yr) + Σ [$H_2$ fuel cost (i) + O&M cost (i) + policy cost (i)] × total number of HFCVs in the fleet (i)

Reference vehicle LCC (i) ($/yr) = # number of new HFCVs (i) × reference vehicle first cost (i) ($/yr) + Σ [gasoline fuel cost (i) + O&M cost (i) + policy cost (i)] × total number of FCVs in the fleet (i)

ΔLCC (i) = reference vehicle LCC (i) ($/yr) – LCC HFCV (i) ($/yr) = number of new HFCVs (i) × [reference vehicle first cost (i) – HFCV first cost (i) ($/yr)] + Σ [gasoline fuel cost (i) – H_2 fuel cost (i) + Δpolicy cost (i)] × total number of HFCVs in the fleet (i)

The difference in life-cycle costs ΔLCC at each year (cash flow) represents the funding that would have to be supplied each year to make the cost of HFCVs equivalent to that of the reference gasoline vehicles. Initially, HFCVs cost a lot more than gasoline vehicles (but the number of new HFCVs is low) so the cash flow is negative. Eventually as costs for HFCVs come down via learning, under some conditions ΔLCC (i) becomes positive.

When the costs are equal, the annual cash flow ΔLCC (i) = 0. The year that this happens is termed the "LCC breakeven" year. Presumably, at this point the net cost to the economy is the same for FCVs and gasoline reference vehicles.

Step 3: Estimating Transition Costs

Add up incremental HFCV vehicle and fuel costs to get to the LCC breakeven year (compared to the gasoline reference vehicle). These are transition or "buydown" costs.

Buydown cost ($) = Σ ΔLCC (i)   i = 1 to the breakeven year

Initially, the first cost of the HFCV will be much higher than that of the reference vehicle. This cost falls over time (with increased learning and mass production of HFCVs), so that eventually, under some conditions ΔLCC (i) = 0, and the negative cash flow "bottoms out."

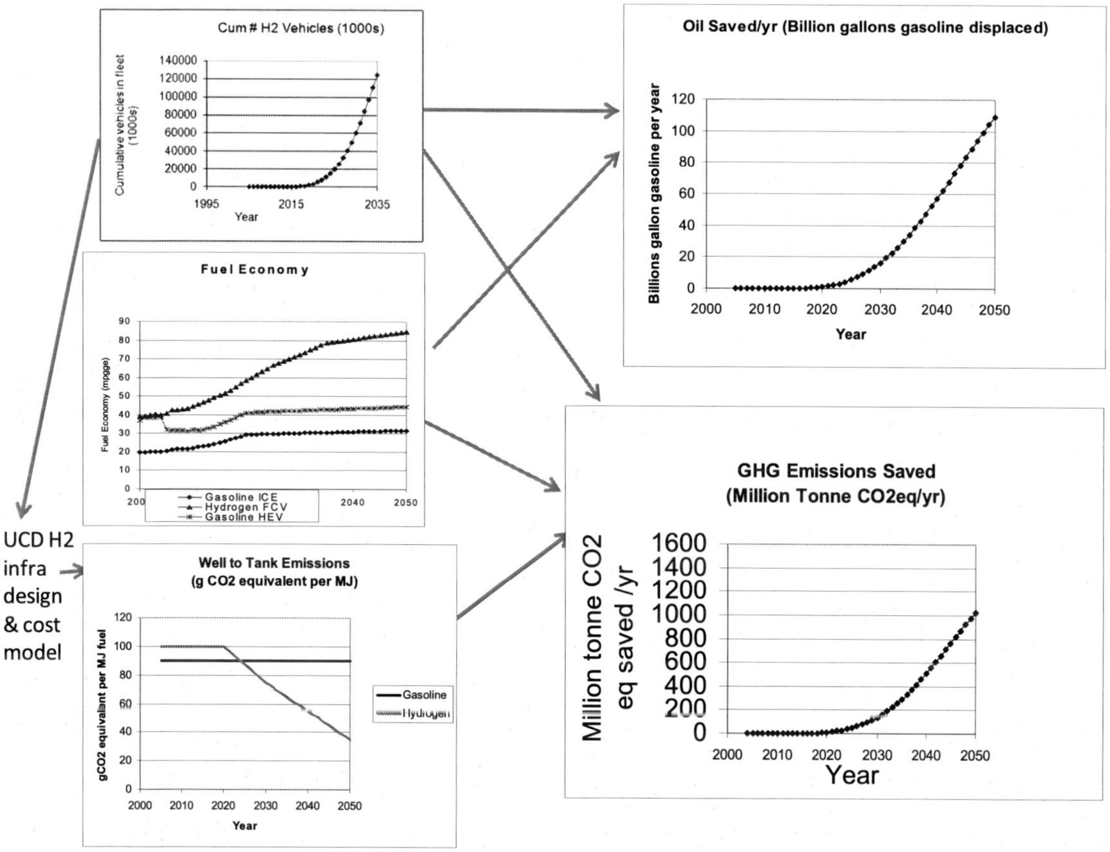

FIGURE C.1(b) Flow diagram of simple transition model (part 2), oil and greenhouse gas emissions saved.

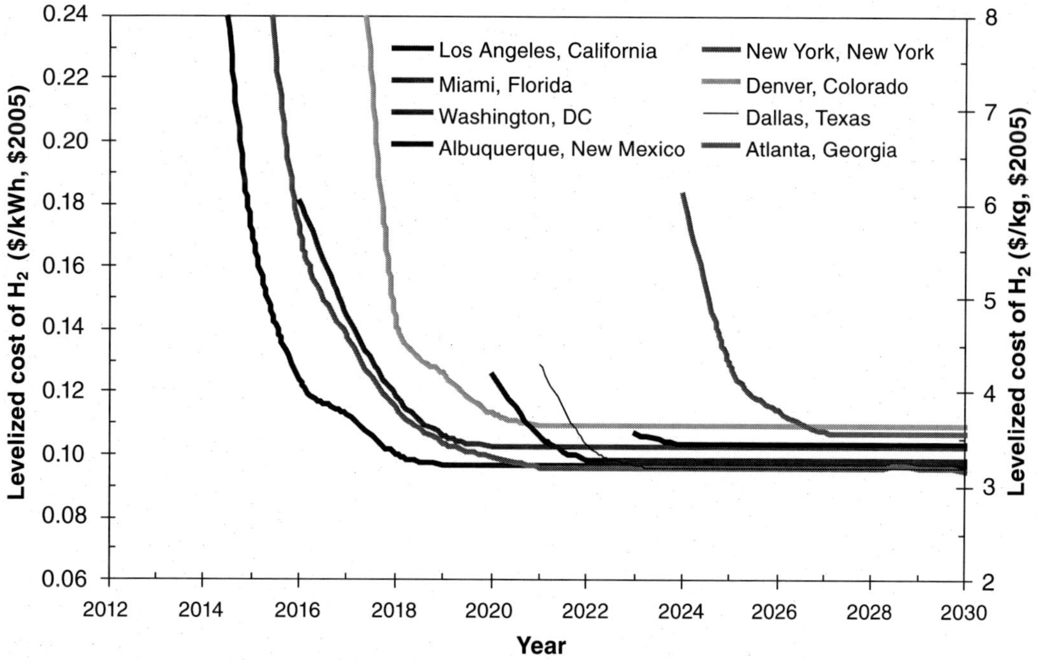

FIGURE C.2 Delivered hydrogen costs in selected cities.

Consider incremental costs for vehicles and H_2 fuel separately:

Incremental vehicle cost ($) = Σ Number of new HFCVs (i) × [first cost HFCV (i) – first cost reference vehicle (i)], i = 1 to the breakeven year

Incremental fuel cost ($) = number of HFCVs in the fleet (i) × [fuel cost HFCV (i) – fuel cost reference vehicle (i)], i = 1 to the breakeven year

Adding up the infrastructure capital costs to the breakeven year gives an indication of cumulative costs to energy companies. These are the cumulative costs that would be borne by automakers or energy companies to reach breakeven.

Step 4: Estimating Policy Costs

- Vehicle subsidy is subtracted from vehicle first cost.
- Fuel subsidy is subtracted from fuel cost.
- Carbon tax is added to operating costs.

Cost for each vehicle becomes:

LCC ($) = (vehicle first cost ($) – vehicle subsidy ($)) + Σ [(fuel costs – fuel subsidy) + O&M costs + carbon emissions × carbon tax)]

The cost of policies can be estimated over time, either to the breakeven year or to some set "policy horizon."

The cost of a direct subsidy to energy providers (e.g., pay for 50 percent of cost of first stations) could be calculated in an analogous fashion.

Step 5: Estimating Savings in Oil Use and GHG emissions (Figures C.3 and C.4)

- Using a vehicle stock model, keep track of the number of HFCVs of each model year in the fleet.
- Each year, the H_2 vehicles displace a certain amount of gasoline use (the gasoline that would have been used by reference gasoline cars, if the HFCVs had not been introduced).
- The HFCVs have certain well-to-wheels GHG emissions, depending on the assumed H_2 supply options (which are estimated separately and input to the scenario). These emissions are lower than those of the reference gasoline vehicle, and GHG emission reductions can be estimated for each year.

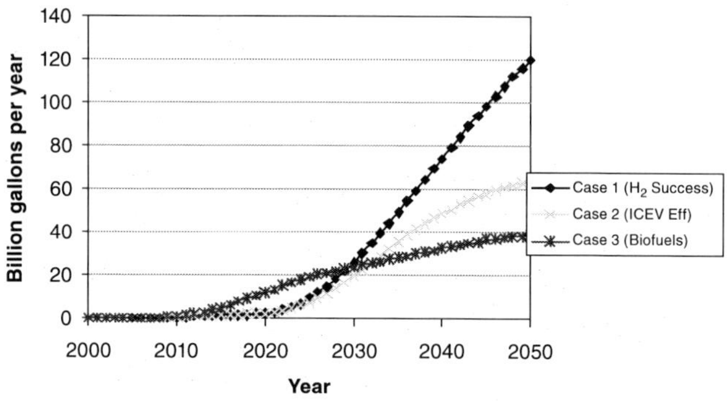

FIGURE C.3 Oil saved per year with different scenarios compared to the reference case.

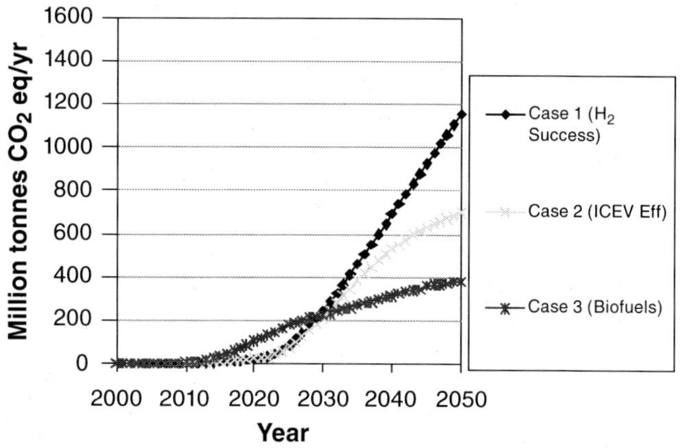

FIGURE C.4 Greenhouse gas emissions avoided compared to the reference case.

Appendix D

Acronyms and Abbreviations

ABE	acetone-butanol-ethanol
AEO	Annual Energy Outlook
AER	all-electric range
C_d	coefficient of drag
CAFE	corporate average fuel economy
CCS	carbon capture and sequestration
CO	carbon monoxide
CO_2	carbon dioxide
CRP	Conservation Reserve Program
CVA	camless valve actuation
CVT	continuously variable transmission
DME	dimethyl ether
DNGR	distributed natural gas reforming
DOE	U.S. Department of Energy
DSMR	distributed steam methane reforming
DWE	distributed water electrolysis
EEG	Renewable Energy Resources Act (Germany)
EIA	Energy Information Administration
EISA	Energy Independence and Security Act of 2007
EPA	U.S. Environmental Protection Agency
EPACT	Energy Policy Act of 2005
EV	electric vehicle
FC	fuel cell
FCFP	FreedomCAR and Fuel Partnership
FCV	fuel cell vehicle
GDI	gasoline direct injection
GHG	greenhouse gas
GIS	geographic information system
GW	gigawatt
HCCI	homogeneous-charge compression ignition
HEV	hybrid electric vehicle
HFC	hydrofluorocarbon
HFCV	hydrogen fuel cell vehicle
HICEV	hydrogen-powered ICEV
ICE	internal combustion engine
ICEV	internal combustion engine vehicle
IEA	International Energy Agency
IGCC	integrated gasification-combined cycle
IPCC	Intergovernmental Panel on Climate Change
ISO	International Organization for Standardization
kWh	kilowatt-hour
LCC	life-cycle cost
LDV	light-duty vehicle
LNG	liquefied natural gas
MCFC	molten carbonate fuel cell
MEA	membrane electrode assembly
mpg	miles per gallon
mpgge	miles per gallon of gasoline equivalent
MPR	maximum practicable rate of penetration
MTG	methanol to gasoline
NEMS	National Energy Modeling System
NO_x	nitrogen oxides
NRC	National Research Council
OEM	original equipment manufacturer
PEM	proton exchange membrane
PFC	perfluorocarbon
PHEV	plug-in hybrid electric vehicle
ppm	parts per million
PTC	production tax credit
PUC	public utility commission
PV	photovoltaic

R&D research and development
RD&D research, development, and demonstration
RPS renewable portfolio standard

SAE Society of Automotive Engineers (now SAE International)
SMR steam methane reformer
SO_2 sulfur dioxide
SOFC solid oxide fuel cell
SUV sport utility vehicle

USDA U.S. Department of Agriculture

VEETC Volumetric Ethanol Excise Tax Credit
VMT vehicle-miles traveled
VVT variable valve timing

WTW well to wheel

ZEV zero emissions vehicle